Fire Protection Engineering

in Building Design

Fire Protection Engineering in Building Design

Jane I. Lataille, P.E.
Fire Protection Engineer
Los Alamos National Laboratory

BUTTERWORTH
HEINEMANN
An imprint of Elsevier Science
Amsterdam Boston London New York Oxford Paris
San Diego San Francisco Singapore Sydney Tokyo

Butterworth–Heinemann is an imprint of Elsevier Science.

Library of Congress Cataloging-in-Publication Data

ISBN: 0-7506-7497-0

British Library Cataloguing-in-Publication Data
A catalogue record for this book is available from the British Library.

The publisher offers special discounts on bulk orders of this book.
For information, please contact:

Manager of Special Sales
Elsevier Science
200 Wheeler Road
Burlington, MA 01803
Tel: 781-313-4700
Fax: 781-313-4882

For information on all Butterworth–Heinemann publications available, contact our World Wide Web home page at: http://www.bh.com

10 9 8 7 6 5 4 3 2 1

Printed in the United States of America

Contents

Foreword

Fire protection is an integral part of building design and must be integrated into the overall design process from the very beginning of the project.

It is vitally important for everyone involved in the building design process—architects; structural, mechanical, and process engineers; interior designers, and other design professionals—to be aware of the fire protection engineering issues that need to be considered at each step in the process.

In this book, Jane Lataille, a well known fire protection engineer with over 27 years of experience in the field, explains in an easy-to-understand, straightforward fashion, what fire protection engineering involves and what issues need to be considered in integrating fire protection into the overall building design process.

This book provides excellent guidance to the non-fire protection engineer on the coordination necessary during the design process to make sure that the fire protection design provides a level of safety acceptable to building owners, insurers, and code enforcers that does not impose unnecessary constraints on the overall building design or operation.

Arthur E. Cote, P.E.
Executive Vice President – NFPA International

Preface

In an ever-tightening economy, protecting assets as economically as possible is highly critical. Fire protection systems protect people, property, and mission, but they can also be expensive. Designing these systems as cost-effectively as possible requires a high level of knowledge about how they work in the built environment.

Older prescriptive-type fire protection codes could sometimes be overly conservative and therefore unnecessarily expensive. Newer prescriptive codes have alleviated some of the inefficiency, but they still might not provide the most effective designs for very specialized buildings.

Performance-based designs allow maximum flexibility while achieving a specified level of protection. With this newfound freedom from prescriptive requirements comes the responsibility for setting goals, selecting appropriate levels of protection, and determining the performance available from the fire protection design options being considered. This requires extensive knowledge of both fire science and fire protection engineering.

Being able to design prescriptive sprinkler or fire alarm systems does not usually constitute a sufficient background for determining fire protection system performance. However, engineers of all disciplines on a project can work with the architect, prime engineering professional, and fire protection engineer to implement performance-based requirements.

The goal of this book is to explain what fire protection engineering involves and how to integrate fire protection design into an overall building project. It describes the coordination between the architectural and engineering disciplines required to accomplish the integration. And it discusses the critical interrelationships between fire protection and building design for both performance-based and prescriptive fire protection criteria.

This book does not explain how to design fire protection systems. It assumes that the fire protection systems on a building project are designed by experienced fire protection engineers with BS degrees or P.E. licenses specifically in fire protection engineering, or by those with comparable training.

The Introduction discusses the importance of integrating fire protection design into the overall building project. The first two chapters lay the groundwork for integrating fire protection design. Chapter 1 reviews what the discipline of fire protection engineering encompasses and where it interfaces with other engineering disciplines. Chapter 2 briefly describes the fire protection systems most commonly used in building projects and the many functions they can serve.

Chapter 3 discusses using performance-based design in meeting fire protection requirements, and explains how this affects all facets of the building design. It stresses the importance of documenting all the factors affecting a performance-based design and of managing future change.

Chapter 4 discusses using prescriptive fire protection design, which is still very common on building projects. Chapter 5 lists areas where fire protection system design interfaces with the traditional engineering disciplines. These interfaces apply to both prescriptive and performance-based designs.

Chapter 6 explains how integrating fire protection design applies to existing buildings as well as to new construction. Chapter 7 addresses writing fire protection specifications, and the References section lists useful fire protection information sources, including professional societies and published references.

The National Fire Protection Association (NFPA) publishes fire codes that architects, engineers, and building officials use every day. However, only the most common NFPA codes are well known. Fire protection is a very complex subject, and so are all the codes that address it. Throughout this book, applicable NFPA codes are cited for each facet of fire protection in buildings.

Even in its better known prescriptive mode, fire protection engineering is often misunderstood or misapplied. Adding performance-based design has made fire protection all the more challenging to grasp. In 2000, The Society of Fire Protection Engineers (SFPE) and NFPA jointly published the benchmark for understanding performance-based fire protection design: *The SFPE Engineering Guide to Performance-Based Fire Protection Analysis and Design of Buildings*. SFPE has also published many articles on performance-based fire protection design in *Fire Protection Engineering* magazine. These sources are indispensable for understanding performance-based fire protection design.

Many people helped this book emerge from its original concept. I would like to thank Morgan J. Hurley, P.E., Technical Director, SFPE; and Brian Meacham, P.E., of Arup Corporation for their review of the book concept and for their insightful comments and suggestions.

Thanks also go to everyone else who reviewed material in this book, including Robert F. Daley, P.E., Morgan J. Hurley, P.E., Brian Meacham, P.E., James R. Streit, P.E., Allen Trujillo, and Julia H. Wood, P.E.

Special thanks go to Arthur Cote, Executive Vice President of NFPA, for writing the Foreword. Finally, I would like to thank the Los Alamos National Laboratory for its support in developing the book.

Introduction: The Importance of Integrating Fire Protection Design

Fire protection is an integral part of the built environment. As such, it should always be engineered in conjunction with the overall building design. Multi-discipline engineering firms sometimes have engineers of other disciplines design the fire protection systems; sometimes they outsource the fire protection design to engineering consultants. Either option can result in inefficiency, improper design, or excess cost if not properly coordinated.

Fire protection design was once almost exclusively prescriptive. In other words, projects incorporated specific fire protection measures prescribed by codes. Prescriptive fire protection design is still commonly used on many projects.

Engineers in disciplines other than fire protection are often charged with designing the fire protection in accordance with prescriptive code requirements. Proper design of fire protection systems for a prescriptive-type project requires coordinating the fire protection design with the overall building design and integrating the fire protection design features with the other engineering disciplines. Fire protection features that are not designed while a building is being planned can sometimes be very difficult to incorporate later. Adding these features later increases the cost; leaving them out compromises the level of protection provided in the building.

In contrast with prescriptive design, performance-based fire protection design considers how fire protection systems perform given the selected building design and its expected fire loading. Performance-based fire protection design is steadily becoming more common. This type of design requires very close coordination with the building design, because every change specified to the building can affect fire protection system performance. Following prescriptive code requirements and coordinating them with the other engineering disciplines is not sufficient.

Just as experienced structural engineers design or oversee the design of bridges, experienced fire protection engineers should design or oversee the design of fire protection systems. Even for prescriptive designs, the information available in codes is not sufficient for a design basis. The fire protection engineer must also understand fire loading, fire development and growth, heat transfer, and how available fire models handle all these elements.

In addition, the fire protection engineer and architect must closely coordinate all fire protection design features and document their place in the performance-based design. For example, if a wall is intended to increase available occupant egress time or to eliminate the need for sprinklers in a particular area, then the interior designer must be made aware that the wall cannot be changed without changing the fire protection design. Many buildings with atria have special design features that likewise should not be changed. Once the performance-based fire protection design features have been selected and documented, they can be specified and coordinated with the other engineering disciplines.

Whether a building is new or existing, or whether the fire protection design is prescriptive or performance-based, this book explains how to integrate fire protection engineering into the building design.

Fire Protection Engineering

in Building Design

1:
What Is Fire Protection Engineering?

1-1 The Discipline

Fire protection engineering is not widely understood by those outside the discipline. Many engineers from other disciplines have never heard of it. Some of them think fire protection engineering is manual firefighting, while others think it is fire code enforcement. Still others think it is forensic engineering (e.g., reconstructing what happened after a fire has occurred). Although fire protection engineering could include elements of any of these activities, it is a far more comprehensive discipline than most people realize.

Fire protection engineering interfaces with all the major disciplines on a building project. From an architectural standpoint, fire protection engineers concern themselves with how building layout affects firefighting access, egress characteristics, and other life safety features.

From a structural standpoint, fire protection engineers concern themselves with the strength, thickness and fire resistance rating of building construction materials; the location of and protection of openings in fire walls or fire barriers; and the ability of a structure to support the weight of water-filled sprinkler piping. They also concern themselves with earthquake resistance.

From a mechanical standpoint, fire protection engineers calculate the flow of water through sprinkler piping, the discharge of special extinguishing agents through nozzles, and flow of air and gases through smoke control systems. From an electrical standpoint, they address the wiring of fire alarm systems, detection systems, special extinguishing systems and fire pumps. They also address backup power supplies, emergency lighting, and electrical equipment for use in hazardous locations.

Finally, from a chemical standpoint, fire protection engineers analyze the hazards of chemical interactions and processes. This includes:

- Recognizing hazards of materials and material interactions;
- Identifying potential sources of ignition;
- Identifying potential sources of spills, amounts that could be spilled, and the consequences of ignition of a spill;
- Determining the consequences of unsafe pressures, temperatures, flows or concentrations of materials in reactions; and
- Analyzing process control systems, including the parameters requiring control, monitoring, interlocks and shutdowns.

Furthermore, fire protection engineers must integrate these diverse building features into a uniform design package.

Like other engineering disciplines, fire protection engineering involves designing devices, systems and processes to serve a particular function. In this case the function is protecting people, property and business operations from the results of fire. Like other engineers, fire protection engineers typically have engineering degrees and might or might not have Professional Engineering (P.E.) licenses.

Fire protection engineering is one of fifteen engineering disciplines that offer a P.E. examination through the National Council of Examiners for Engineering and Surveying (NCEES). (See References.) NCEES publishes several sources of information about fire protection engineering, including an exam syllabus and a standard of minimal competence.

The P.E. examination must cover all the subjects on the fire protection exam syllabus. These subjects illustrate what the discipline encompasses. (See Figure 1.)

Figure 1: Subjects from the NCEES P.E. Exam Syllabus for Fire Protection Engineering

PLANNING AND DESIGN OF WATER SUPPLIES
Water supplies dedicated to fire protection, public water supplies

PLANNING AND DESIGN OF BUILDING SYSTEMS
Structural fire resistance, fire barriers, opening protection, means of egress, construction materials, smoke management systems, building use and occupancy

PLANNING AND DESIGN OF WATER-BASED SUPPRESSION SYSTEMS
Specifying, evaluating, testing, and maintaining sprinkler and water spray systems; fire and explosion suppression systems

PLANNING AND DESIGN OF NONWATER-BASED SUPPRESSION SYSTEMS
Specifying, evaluating, testing, and maintaining CO_2, dry chemical, foam, and alternate agent systems; fire and explosion suppression systems

PLANNING AND DESIGN OF DETECTION AND ALARM SYSTEMS
Specifying, evaluating, testing, and maintaining heat, smoke and flame detectors; alarm and supervisory systems

PLANNING AND DESIGN OF FIRE PREVENTION
Control of combustible materials, ignition sources, and oxidizing agents

IMPLEMENTATION AND MONITORING OF FIRE PREVENTION
Inspection, testing, and preventive maintenance; process safety; hazard abatement

RESEARCH AND DEVELOPMENT OF HAZARD AND RISK ANALYSIS
Quantification of frequency and severity of fire events, estimation of time available for occupant egress from rooms, analysis of damage potential to exposed objects from fire or explosion

As can be seen from this syllabus, fire protection engineering encompasses facets from all the major engineering disciplines: structural, mechanical, electrical, and chemical engineering. These facets of fire protection engineering must be addressed as a system for them to work together properly in a building. The ability to integrate these wide-ranging facets into an effective design is one of the greatest strengths of the fire protection engineering discipline.

In addition to the exam syllabus, NCEES also publishes a Standard of Minimal Competence for each engineering discipline. This standard briefly describes what minimally competent engineers are expected to understand. It is used to find the appropriate difficulty level of P.E. examination problems. Figure 2 reproduces the Standard of Minimal Competence for fire protection engineers.

Figure 2: NCEES Standard of Minimal Competence for Fire Protection Engineers

The minimally competent Fire Protection Engineer must possess:

- A thorough understanding of fundamental fire protection systems and practices as they pertain to life safety and to fire prevention, detection, control, and extinguishment. This includes the ability to apply this understanding in conjunction with commonly used fire protection standards;
- A working knowledge of the nature and characteristics of fire and related hazards, including how fires originate, develop, and spread;
- A basic understanding of the effects of fire and fire protection measures on life, property, operations, and the environment;
- A basic understanding of hazard and risk; and
- An awareness of related fire protection standards and tools.

1-2 The Professional Society

Another good source of information about fire protection engineering is the Society of Fire Protection Engineers (SFPE). (See References.) As the primary professional society for fire protection engineers, SFPE is concerned with what fire protection engineering encompasses and the qualifications of those practicing it.

SFPE defines fire protection engineering as follows:

> Fire Protection Engineering is the application of science and engineering principles to protect people and their environment from destructive fire and includes:
>
> 1. analysis of fire hazards;
> 2. mitigation of fire damage by proper design, construction, arrangement, and use of buildings, materials, structures, industrial processes, and transportation systems;
> 3. design, installation, and maintenance of fire detection, suppression and communication systems; and
> 4. post-fire investigation and analysis.

SFPE also defines a Fire Protection Engineer:

> A Fire Protection Engineer (FPE) by education, training, and experience:
>
> 1. is familiar with the nature and characteristics of fire and the associated products of combustion;
> 2. understands how fires originate, spread within and outside of buildings/structures, and can be detected, controlled, and extinguished; and
> 3. can anticipate the behavior of materials, apparatus, and processes as related to the protection of life and property from fire.

These definitions track with both the P.E. standard of minimal competence and with SFPE membership requirements.

1-3 What FPEs Do

Most FPEs do not work in all the categories listed on the P.E. exam syllabus. A typical FPE works in several fields falling under one or more of these categories. For example, a suppression system designer might evaluate the hazard to be protected, select detection methods, specify suppression system performance, and lay out the system. Or a fire protection consultant might conduct hazard analyses and compare the overall risk to an entire facility from various combinations of fire protection design options.

The underlying requirement is that FPEs be qualified by experience and training in their work areas. This is true whether or not the FPE has a degree in fire protection engineering, a degree in another engineering field, or a P.E. license.

FPEs responsibilities vary with their employer. Employers of FPEs include:

- Consulting firms;
- Educational institutions;
- Fire protection associations and societies;
- Fire protection equipment manufacturers;
- Fire testing laboratories;
- Government agencies;
- Industry;
- Insurance companies; and
- Municipalities.

Employers involved in building design need the most comprehensive understanding of how fire protection engineering interfaces with the other engineering disciplines. This is one reason why they hire fire protection engineers.

Many job functions in fire protection-related fields do not fall directly in the P.E. exam categories, but they can still interface with many facets of building system design. Such job functions include:

- Alarm/detection system technicians;
- Building officials;
- Emergency response teams;
- Extinguishing system technicians;
- Fire marshals;
- Fire protection system plan reviewers;
- Fire science researchers;
- Forensic investigators;
- Hazard evaluators;
- Industrial fire protection/security officers;
- Insurance company fire protection representatives;
- Life safety professionals;
- Process safety systems technicians; and
- Sprinkler system technicians.

As an example, the responsibilities of a sprinkler system technician could include laying out sprinkler systems in accordance with engineering specifications or confirming that a given sprinkler system layout meets a specified design. Personnel in these related fields are rarely responsible for coordinating fire protection with other disciplines, though they may be aware of the interrelationships.

1-4 How Fire Protection Engineering Differs

Few practitioners of the major engineering disciplines have an in-depth knowledge of fire protection engineering. This is because the major disciplines apply engineering concepts to certain traditional design areas. For example, mechanical engineers apply the concepts of fluid flow to design plumbing and HVAC systems. This works well because plumbing and HVAC system loads are usually easy to determine.

The potential problem with sprinkler system design is that there is much more uncertainty about the potential heat load (i.e., what the sprinkler system hydraulic design should be). In addition, different reliability and maintenance considerations apply to sprinkler systems because they are primarily idle, while other mechanical systems are in constant use. Mechanical engineers are not normally trained in how to handle these considerations.

This is just one example of how knowing what fire protection engineering encompasses can help integrate it in a building project. Later chapters give many other examples.

For additional information on the discipline of fire protection engineering, see *Fire Protection Engineering* magazine, Issue Number 3 (Summer 1999). This issue, subtitled "Progress in Professional Practice," contains four articles about different facets of the discipline.

2:
Functions of Fire Protection Systems

2-1 Preventing and Protecting Against Fire

Having an adequate level of protection against fire is important in meeting facility goals. However, preventing as many fires as possible is just as important, if not more so. Preventing fires is accomplished through a facility's fire prevention programs.

The fire prevention measures based on engineered systems must be implemented in the project design stage. In this respect, fire prevention and fire protection measures closely overlap. Sometimes no distinction is drawn between them. Engineered fire prevention measures can include:

- Separation distances between hazards and exposures;
- Combustion safeguards on fuel-fired equipment;
- Systems for liquid containment, drainage or run-off;
- Provisions for bonding and grounding to control static;
- Explosion-proof electrical and heating equipment in hazardous areas; and
- Process safety control systems.

Fire prevention measures based on programs and procedures (as opposed to engineered systems) are not often considered in the planning stages of construction, despite the fact that this is the best time to develop them. The fire protection engineer generally recommends appropriate fire prevention programs for each project. For these programs to be effective, the project team must help integrate them into the project design.

Fire protection systems are of many types. Selecting the appropriate type requires understanding the hazard to be protected, the types of protective systems that are appropriate for that hazard, and the level of protection each type of system can be expected to provide.

Examples of different types of fire protection systems include:

- Detection systems with interlocks for door or damper closure, HVAC shutdown, or process shutdown;
- Fireproofing for buildings, structures, or processes;
- Fire walls, fire barriers, fire doors, and other fire resistant construction;
- Inerting systems;
- Smoke control systems;
- Sprinkler systems;
- Deluge and preaction systems; and
- Special extinguishing systems, including those using wet or dry chemicals, foam, or "clean" agents.

Whether a design is prescriptive or performance-based, understanding of the following elements is essential for proper fire protection design:

- Reason(s) for installing the system;
- Assets being protected;
- Function the system is serving; and
- Science behind the system design.

The remainder of this chapter addresses the first three elements. Chapters 3 and 4 address the fourth element.

The discussion of fire protection systems in this chapter assumes that appropriate fire prevention programs are already in place or are being planned. The subject of fire prevention programs is beyond the scope of this book. Many existing books address this subject in great detail.

2-2 Reasons for Installing Fire Protection Systems

Fire protection systems can be installed for many different reasons. Most often, fire protection systems are expected to meet a combination of purposes. Designing a fire protection system requires knowing the purposes it must serve.

Requirements to install fire protection systems usually stem from mandatory codes, but the systems installed to meet these codes will not necessarily meet all the owner's goals unless this is specified.

Reasons for installing fire protection include the following:

Meeting codes. Most fire protection systems are installed to meet codes. In the U.S. this means NFPA 13 as well as other NFPA codes. The U.S. regional building codes also require installing fire protection systems.

Making trade-offs. Sometimes installing additional fire protection allows more flexibility in architectural design. For example, installing curtain water spray systems might allow having an open atrium in a mall.

Satisfying AHJs. Based on conditions in a particular jurisdiction or in a particular building, an AHJ could require fire protection systems that are not addressed in the applicable codes.

Protecting assets. Fire protection systems can be installed to protect a building or a building's contents, to control specific hazardous processes or areas, to safeguard human life, or to preserve mission continuity. The level of fire protection required for protecting particular assets can sometimes exceed the minimum required by codes.

Maintaining community relations. Sometimes an isolated, small-valued hazard that would not normally require or warrant fire protection is protected for the good of the community. One example of this is protecting a hazard that has the potential for causing damage to neighboring properties.

Most fire protection systems are installed for several of the above reasons. One of the challenges of designing fire protection systems is to achieve several purposes as effectively as possible. Another challenge is to anticipate likely future occupancy changes in the original fire protection design basis.

Chances that fire protection systems will serve a building's needs are greatly increased if they are coordinated throughout the project. A good reference for coordinating building code needs is *Cracking the Codes*, by Barry Yatt (see References). Chapter 5 of this book addresses coordinating with fire protection-related codes. Similar coordination is also needed for noncode needs. The building owner must coordinate these needs by working with the project team.

2-3 Protecting Assets

Asset protection is a very important function of fire protection systems. Assets that fire protection systems can be intended to protect include:

Property. Conventional sprinkler systems protect buildings. In-rack sprinkler systems keep fire from spreading through rack storage. Sprinkler systems limit property damage, but they cannot totally eliminate it. Directional water spray systems protect special hazards, like oil-filled transformers. Protecting a transformer does not save it from damage, but keeps it from damaging nearby buildings and structures, including other transformers.

Special extinguishing systems, such as those using gaseous agents, are sometimes used to protect critical computer or data processing facilities. These extinguishing systems are designed to actuate before conventional sprinkler systems would actuate, and they can extinguish fire while damage is still minimal, even preventing some equipment damage. Sprinkler systems are still provided as back-up protection for the building.

On the other hand, explosion suppression systems can protect equipment and structures from damage. These systems operate so fast that the pressure wave started by ignition of an explosive atmosphere is suppressed before it reaches a high enough pressure to cause any damage.

Life. Controlling fire sufficiently to protect a building can also keep fire from harming people. Since people are also harmed by the smoke fire generates, smoke control systems are used to allow time for people to evacuate before smoke concentrations reach dangerous levels.

The basis for protecting life is in ensuring fast egress from buildings. This involves:

- Provision of adequate exit capacity;
- Maximum allowed distances for egress travel paths;
- Minimum allowed widths of egress travel paths;
- Reliably illuminated and marked exits;
- Maximum allowed length of dead ends; and
- Protected exits to public ways.

All the above features depend on the number of occupants in a building and their mobility. NFPA 101, *Life Safety Code,*© and model building codes address these features.

Mission continuity. After a fire, lost property can be replaced and damaged buildings can be repaired. But business lost to competitors while operations are down cannot always be recovered. Competitive industries sometimes provide more fire protection than required for protection of life and property to decrease possible downtime that may occur after a fire.

Protecting mission continuity requires not only carefully designed fire protection systems, but also effective fire prevention programs. Engineers who only design fire protection systems may not know what fire prevention programs are necessary. Fire protection engineers are usually very familiar with developing these programs.

Environment. Risk management principles often dictate protecting lives and high value property. Unoccupied buildings of relatively low value may not normally require protection. However, this changes if a fire in such buildings could have an adverse effect on the environment. This could be due to the contents of the building or to its location near a waterway or watershed area.

Protecting the environment boils down to asset protection for two reasons. First, a company could be held liable for environmental damage caused by a fire on its property. Second, an unpolluted environment is everyone's asset.

2-4 Relating Design Features to Function

Knowing the function of fire protection systems to be installed and what they are expected to protect is essential for designing them properly. Fire protection system design takes many functions into account:

Detection. A common misconception is that fire detection is a form of protection. Some might argue that a building with smoke detectors does not need sprinklers. This is not true. Fire protection systems may require detection to operate, but detection alone does not constitute protection.

Note that in cases where risk analysis has determined that a fire protection system need not be provided, detection can be provided for other reasons. These reasons may include process shutdown or occupant notification.

Given that detectors actuate fire protection systems, the optimum detector type should be chosen. Conventional sprinklers operate as heat detectors and are suitable for protecting ordinary combustibles. Smoke detectors operate smoke control systems. Special extinguishing systems can be operated by any type of detector. The type of detector is selected to match the hazard being protected.

Available types of detectors include the following:

- Conventional spot-type ionization and photoelectric smoke detectors;
- Duct-type smoke detectors;
- Line-type photoelectric smoke detectors;
- High sensitivity spot-type smoke detectors;
- High sensitivity air sampling smoke detectors;
- Fixed temperature heat detectors, including sprinklers;
- Rate-of-rise heat detectors;
- Rate-of-rise compensated heat detectors;
- Flame detectors;
- Pressure sensors for sensing air shock waves generated in the early stages of a deflagration;
- Combustible gas sensors;
- Oxygen sensors;
- Sensors for temperatures, pressures, flows, liquid levels, and other process parameters;
- Sensors for detecting presence of liquids; and
- Position limit switches.

Occupant warning. The time occupants have to evacuate a building depends on how soon they are notified of conditions requiring evacuation. The detection system or systems used determine how promptly occupants are notified.

curtain installed for exposure protection. Water curtains can spray on outside building walls to protect a building from an external fire exposure, or they can spray on glass walls facing an atrium inside a building. They can be used in many other ways. Protecting a building, structure, or process from fire in an exposing hazard does not mean that the exposure itself need not be protected. This issue must be considered independently.

Fire control. This is the most common goal of the familiar sprinkler system. Code-compliant sprinkler systems are designed to control fire, but not necessarily extinguish it. Final extinguishment usually depends on fire department operations or other manual intervention. A facility's risk analysis needs to take this into account. In other words, the analysis should not assume that sprinkler systems extinguish any fire completely.

In some areas, fire extinguishment by an automatic fire protection system may sometimes be desirable. Examples are inaccessible areas or areas that might be too dangerous for people to enter. Different types of fire protection systems or fire protection system design can accommodate this need.

Fire extinguishment. In enclosed areas, properly designed total flooding gaseous extinguishing systems can extinguish fire. In storage buildings, properly designed sprinkler systems using ESFR (Early Suppression Fast Response) sprinklers can extinguish fire. Systems using ESFR heads have many stringent design rules, and even small deviations from these rules can render the systems ineffective.

Other types of systems that can extinguish fire include inerting systems, spark suppression systems, and explosion suppression systems. Other methods that can extinguish fire include interlocks that automatically drop lids over open tanks when smoke, heat, or fire is sensed. Fire protection engineers can design control and extinguishment systems for many types of hazards.

3:
Performance-Based Fire Protection Design

3-1 Design Elements

Engineers in the major disciplines commonly use performance-based designs. Structural engineers design bridges to withstand a particular load. Mechanical engineers design air conditioning systems to cool an area by a given number of degrees in a specified time. Two elements are required to make performance-based design possible:

1. *The underlying science must be well understood and developed.* In the case of bridge design, the physics of structural loading is contained in the Newtonian equations for balancing forces. In the case of cooling system design, the thermodynamic properties of fluids are embodied in heat transfer equations.

2. *The design loads must be known.* Maximum traffic loads can be set for a bridge, and snow, wind, and earthquake loads are obtained from codes that are based on historical information. The maximum amount of cooling required for a building can be determined from local climate information, the location and number of windows, and the amount of heat expected to be generated by equipment and occupants.

Twenty years ago, the underlying science of fire protection engineering—called fire science, or fire dynamics—was in its infancy. It was not well enough developed to serve as a basis for performance-based designs. Fire science has since been much more highly developed. In theory, it can now be used to calculate the results of any fire scenario. In practice, it is used mainly for simple scenarios, because of the extensive amount of calculations required for the more complex scenarios tax the capacity of today's computers. The challenge is to use the simple scenarios as realistically as possible. This requires a thorough understanding of the models now available to fire protection engineers.

Determining realistic fire loads also involves many challenges. The possible arrangements of fire loads in most buildings are so numerous that no design could account for all of them. Fire protection engineers often address this difficulty by determining worst case fire loads, or bounding loads. Sometimes fire protection engineers determine the most likely fire loads for many different scenarios and analyze them all. The potential problem with using the most likely fire loads is that relatively minor changes to a building can result in requiring a new analysis and additional fire protection, unless the original analysis was sufficiently conservative.

The assumed fire loads and design fire scenarios must then be documented. Whenever any feature or use of the building departs from the documented assumptions, the performance-based design may no longer be valid. Selecting appropriate fire loads and design scenarios is therefore extremely critical to the performance-based design process.

Understanding the science and being able to determine fire loads is just the beginning. To implement a performance-based design, the applicable code must permit it—either by being a performance-based code or by allowing performance-based alternatives to prescriptive code provisions. If such designs are permitted, performance criteria must be agreed upon, plausible designs must be developed, the designs must be tested against the performance criteria, and a final design must be selected. Other considerations include coordinating the design with the other disciplines, developing and updating the design documentation, and getting the authority having jurisdiction to accept the design.

The SFPE Engineering Guide to Performance-Based Fire Protection Analysis and Design of Buildings, published jointly by the

National Fire Protection Association and the Society of Fire Protection Engineers in 2000, provides detailed and helpful guidance in implementing performance-based design projects. As the title implies, this guidance can be used to analyze existing buildings or to design new ones.

The Guide presents a process for performance-based design centered around the following major stcps:

1. Defining the Project Scope
2. Identifying the Fire Safety Goals
3. Defining Stakeholder and Design Objectives
4. Developing Performance Criteria
5. Developing Design Fire Scenarios
6. Developing Trial Designs
7. Evaluating Trial Designs
8. Selecting the Final Design

Each step in this process requires an understanding of:

- Fire hazards and risk;
- Characteristics of fire;
- How fires start, develop, and spread;
- How fires affect people, buildings, and processes;
- The underlying science of any fire models used; and
- Principles of fire prevention, detection, and control.

These subjects are included in fire protection engineering curricula and are tested on the Fire Protection P.E. Exam. Professional fire protection engineers draw on their knowledge of these subjects to accomplish the steps in the performance-based design process.

The SFPE guide also covers the written reports required to properly document performance-based design projects and what elements they should contain.

The performance of existing buildings can be analyzed when any changes are being planned. Performance-based analysis is particularly useful when it would be difficult to bring an existing building into compliance with prescriptive fire protection requirements. This is becoming a very common way of handling changes to existing buildings.

The next three sections of this chapter discuss fire science (the underlying science), design fire scenarios (design loads), and other considerations in performance-based fire protection design. The last section gives examples of projects having performance-based fire protection designs.

3-2 Fire Science

Fire science applies the principles of thermodynamics and fluid mechanics to calculate various characteristics of diffusion flames. For example, several flame height correlations have been developed that express flame height as a function of Froude number and the size of the burning surface. Each correlation applies to a particular range of Froude numbers. Using these correlations properly requires knowing enough about the fuel to make a reasonable determination of the Froude number.

Many simpler flame correlations have been developed that depend only on the heat release rate of the fuel. These correlations were developed for particular fuels and/or fuel configurations, and some of them were developed to fit empirical results. Like using the more complex flame height correlations, using the simpler ones requires knowing when they are suitable. It also requires knowing how to determine a reasonable effective diameter of the flame source. The more irregular the source, the harder the effective diameter is to determine.

Because the actual height of a flame varies constantly, calculated flame height must be considered a statistical quantity. The flame height correlations described above calculate mean flame height. Variation from the mean height must also be considered.

Fire science correlations have also been developed for fire plume temperatures and velocities. These correlations are derived from conservation laws using assumptions about gas buoyancy and air entrainment by the plume. Their final forms depend on many factors, including gas density variations and flame height-to-diameter ratio. Like flame height, plume temperatures and velocities must be considered statistical quantities.

Other relevant characteristics of fire that can be calculated are heat release, heat transfer to exposed surfaces, and ignition of exposed surfaces. These calculations develop the initial flame into a fire scenario. Although the calculated fire is still smaller than the real world fires of concern to engineers, it forms the basis for calculating larger fires. It's similar to calculating the structural force on one bridge support before putting together the whole bridge.

Many fire protection references give the equations for calculating the fire characteristics described above, as well as the assumptions on which these equations are based. These references include *The SFPE Handbook of Fire Protection Engineering*, *The NFPA Fire Protection Handbook*, and *An Introduction to Fire Dynamics*, among others. Many of these equations are also given in more general engineering handbooks, such as *Marks'* or Perry's *Chemical Engineers' Handbook*. (See References.)

Fire models repetitively calculate the equations for small flames over larger areas and times. They don't model fire so much as its effects on the compartment in which it is burning. Knowing the effects of fire is sufficient for analyzing a performance-based design. However, the analysis is only as good as the characteristics of the fire selected for modeling.

The effects of fire include temperature increase, smoke buildup, and flashover. Fire models estimate how a fire increases the temperature in a compartment and fills the compartment with smoke. Some of these models account for the different effects when a fire reaches compartment walls and corners.

Examples of the many effects that fire models estimate are:

- Temperatures of fire plume, fire jet, smoke layer, and lower compartment;
- Plume velocity;
- Height of smoke layer;
- Time to flashover;
- Ventilation limits;
- Mass flow through openings and vents;
- Time to ignition of a target;
- Flame spread;
- Sprinkler/detector actuation;
- Fire endurance of structural materials;
- Smoke travel; and
- Occupant egress.

Applying fire models appropriately requires knowing which effects they estimate, what approximations they make, what limitations apply, and how the results affect the risk of the facility being designed or modified.

References useful for understanding fire, fire effects, and appropriate uses of fire models include:

- *An Introduction to Fire Dynamics*, by Dougal Drysdale
- *Principles of Fire Behavior*, by James G. Quintiere
- *Enclosure Fire Dynamics*, by Bjorn Karlsson and James G. Quintiere
- *The SFPE Handbook of Fire Protection Engineering*

3-3 Design Fire Scenarios

In selecting design fire scenarios (the design fire load) for a performance-based design, all possible fire scenarios should be considered. Determining all possible fire scenarios requires knowing as much as possible about the building and its contents and occupants.

Examples of necessary building information are its construction, layout, and building services. Relevant features include fire resistance ratings, fire cutoffs, and the type and arrangement of building services (electricity, gas, oil, HVAC, communications, etc.). Information about existing or proposed fire protection systems would also be relevant. Obtaining information about the building is usually fairly straightforward.

Examples of information needed about building contents are processes, operational characteristics and combustible loading. Relevant features include hazardous materials used in the processes, process energy input and output, process material flow, and the likelihood of the occupancy to change. In most buildings, the processes and operational characteristics dictate the combustible loading. Determining the likely combustible loading can be very challenging, but it is one of the most important factors in estimating reasonable fire characteristics.

Necessary information about occupants includes their number, distribution throughout the building, familiarity with the building, and physical and mental capabilities. This enables a performance-based design to account for and control the effects of fire on people.

Many resources are available for identifying possible fire scenarios. Historical data about the facility and about facilities of similar occupancy can be useful. Simple brainstorming about "what if?" an event occurs can also yield useful results. More analytical techniques include event tree analysis, fault tree analysis, failure analysis, and hazard operability studies.

Once all possible fire scenarios have been developed, they can be sorted into groups with similar likelihood and outcomes. The next step is then to select a representative fire scenario from each group with risk that exceeds the agreed upon level of acceptable risk. The filtered set of scenarios form the basis of the design fire scenarios to be used in the performance-based design.

In selecting the design fire scenarios, the fire protection engineer must be sure that they bound the potential hazards. The most likely fires could be too lenient; likewise, the worst possible case can be too severe. The selected design fire scenarios should reflect the facility's risk from fire as accurately as possible while being appropriately conservative.

3-4 Other Design Considerations

No matter how suitable a performance-based design may be, it can only be implemented if allowed by the applicable codes. Today's prescriptive codes are moving toward accepting performance-based design alternatives as equivalent to meeting particular prescriptive code provisions. In the absence of performance-based alternatives in the codes, some authorities having jurisdiction will accept such equivalencies.

The next logical step is for the codes themselves to be performance-based. Instead of specifying how to design fire protection systems, they will specify the performance criteria these systems are required to meet. The fire protection engineer would then be responsible for developing a design and showing that it meets the performance criteria.

Whether for a performance-based equivalency or a performance-based code, developing performance criteria involves determining an acceptable level of risk. Determining this level can be a problem because of the widespread, but erroneous, belief that following a prescriptive code reduces the risk to zero. Reluctance to document an acceptable level of risk can hold up acceptance of a performance-based design philosophy. Resolving issues like this requires increasing people's understanding and awareness of risk.

Given that a performance-based design has been accepted, all the assumptions on which the design is based must be documented. Performance-based fire protection designs usually make assumptions about the building construction and layout, utility systems, use and occupancy, combustible loading, and occupants. Future change to any of these features has the potential to affect the validity of the fire protection design. Therefore, a good performance-based design should account for the changes most likely to be made in a building.

Just as structural engineers are best equipped to develop performance-based structural designs for bridges, fire protection engineers are best equipped to develop performance-based fire protection designs for buildings and other projects. Many new fire protection hardware and software tools are developed each year, and a thorough understanding of these tools is essential for using them effectively.

The article "The New Toolbox for Fire Protection Engineers," by J. Kenneth Richardson, discusses this concept. This article appeared in the premier issue (Winter 1999) of *Fire Protection Engineering* magazine, a publication of The Society of Fire Protection Engineers. The following quote from this article explains:

> Knowledge of how to approach a problem, how to select and use a model, what's an appropriate input, and how to interpret the calculated results only comes from education and experience. Professional knowledge—what's often called "engineering judgement"—is essential.

Fire protection engineers use their engineering judgement to develop effective performance-based fire protection designs. Those not trained in the discipline can sometimes be misled by popular myths. See "Shattering the Myths of Fire Protection Engineering," also in the premier issue of *Fire Protection Engineering* magazine.

A further complicating factor in the case of fire protection design is that it interfaces intimately with the other engineering disciplines on a building project. Therefore, practitioners of the other disciplines need to be aware of these interrelationships. This chapter explains how the design fire scenarios relate to the other disciplines. The next two chapters address additional areas where the disciplines interface.

3-5 Examples of Performance-Based Design

Performance-based fire protection design has been found to be well suited for special-purpose buildings, such as enclosed sports arenas and other large places of public assembly. One reason for this is that meeting the provisions in prescriptive codes may not always considered the most effective way of protecting these buildings. Another reason is that the future use of these buildings is unlikely to change. Assumptions critical to the performance-based design are therefore likely to remain valid over the life of the building.

A long-held tenet of prescriptive codes is that every story of a building must be cut off from the other stories by construction with a given fire resistance rating. Performance-based design can accommodate vertical openings while meeting all applicable fire protection performance criteria. Likewise, performance-based design can accommodate many other features not normally contemplated in prescriptive codes. The key is in developing appropriate performance criteria and verifying that the design meets them.

Fire Protection Engineering magazine has published many general articles about performance-based design, as well as many case studies about projects using performance-based design. The following summaries give an idea of the range of these articles.

General Articles

Issue Number 7 (Summer 2000):

"Using Models to Support Smoke Management System Design," by James A. Milke, Ph.D. Zone, field and network flow models can help validate smoke control system design.

Issue Number 7 (Summer 2000):

"An Overview of Atrium Smoke Management," by John H. Klote, Ph.D., P.E. Smoke management system designs are based on several computer models.

Issue Number 8 (Fall 2000):

"Pathfinder: A Computer-Based, Timed Egress Simulation," by Joe Cappucio, P.E. This egress simulation computer software tracks individuals and tracks evacuation by room or floor. It can be coupled with a fire model to form a portion of a fire hazard analysis.

Issue Number 10 (Spring 2001):

The entire issue is devoted to managing risk in fire protection. All five articles address facets of how performance-based design addresses risk.

Issue Number 11 (Summer 2001):

"Performance Metrics for Fire Detection," by John M. Cholin, P.E., and Chris Marrion, P.E. To keep up with performance-based design methods, improved prediction of detector response is needed. A possible solution for smoke detectors is to have one metric for detector sensitivity and another for smoke entry delay.

Issue Number 12 (Fall 2001):

"Proactive vs. Prescriptive Fire Protection for the Offshore Industry," by John A. Alderman, P.E., CSP, and Marion Harding, P.E. The offshore industry, which has typically used prescriptive design, is better matching fire protection to risk through performance-based design.

Issue Number 13 (Winter 2001):

The entire issue is devoted to fire models; including how to evaluate, select, and apply them; and models currently being developed.

Issue Number 14 (Spring 2002):

"The Performance-Based Design Review Process Used in the City of Phoenix," by Joe McElvaney, P.E. How the city of Phoenix handles performance-based designs, including an example.

Issue Number 14 (Spring 2002):

"Applications of the Fire Dynamics Simulator in Fire Protection Engineering Consulting," by Jason Sutula. Desktop computers run NIST's CFD models on a variety of scenarios.

Case Studies for Existing Buildings

Premier Issue (Winter 1999):

"Fire Protection for the Star Spangled Banner," by Michael J. Rzeznik, P.E. Protecting the fragile fibers in this historic flag required a specially designed combination of fire prevention and fire suppression systems.

Issue Number 2 (Spring 1999):

"Rehabilitating Existing Buildings," by John M. Watts, Jr., Ph.D. Performance-based design can preserve the character of historic buildings while providing a level of protection comparable to the current code.

Issue Number 5 (Winter 2000):

"Small Atria Smoke Control," by Kurt Ruchala, P.E. A performance-based code equivalency meets the needs of this renovated three-story college dormitory.

Issue Number 8 (Fall 2000):

"Performance-Based Analysis of an Historic Museum," by Andrew Bowman. Renovation of an important historic museum required performance-based analysis of fire safety.

Issue Number 14 (Spring 2002):

"Smoke Control Analysis of a High-Rise Building Using a Network Flow Model," by Sanjay Aggarwal, P.E., Brian D. Gagnon, and Mark D. Reed, P.E. A network flow model analyzes smoke travel in a 14-story building.

Issue Number 14 (Spring 2002):

"Application of a Systematic Fire Safety Evaluation Procedure in the Protection of Historic Property," by Alexander G. Copping, Ph D. A systematic evaluation procedure is applied to historic churches.

Case Studies for New Buildings

Issue Number 3 (Summer 1999):

"A Smoke Management Analysis of a Regional Performing Arts Center," by Eric Rosenbaum, P.E., Scott Laramee, and Craig Beyler, Ph.D. A smoke exhaust system is shown to satisfy the occupant safety objectives of a new performing arts center.

Issue Number 4 (Fall 1999):

"Performance-Based Structural Fire Safety for Eiffel Tower II," by Edward Fixen, P.E. A performance-based code equivalency allows this replica to duplicate the exposed construction of the original Eiffel Tower.

Issue Number 5 (Winter 2000):

"Performance-Based Design of a Professional Hockey Arena," by Michael A. O'Hara, P.E. and Ryan Bierwerth. A combination of prescriptive and performance-based design protects occupants from smoke in this sports arena.

Issue Number 6 (Spring 2000):

"A Risk-Based Fire-Engineered Alternative for Nursing Homes," by Tony Parkes and Carol Caldwell, P.E. Reliability and performance of sprinklers and smoke detectors replace the usual code-required self-closing fire doors to the rooms.

Issue Number 12 (Fall 2001):

"Fire Safety Design of the Fundación Caixa Galicia Building in Spain," by George Faller, C.Eng. Performance-based design helped make this cultural center a work of art.

4:
Prescriptive Fire Protection Design

4-1 Desirability of Prescriptive Design

Despite the advent of performance-based design, much fire protection design is still prescriptive. An important advantage of prescriptive design is that it requires little analysis, and therefore (presumably) little time or knowledge to apply. Implementing prescriptive design is very much like following a recipe.

Another advantage of prescriptive design is that it can cover a broad range of conditions. This is appropriate given the diversity of facilities being protected and the wide-ranging properties of fire. Through its inherent safety factors, prescriptive design can sometimes be more flexible than custom performance-based design.

Many other factors have kept prescriptive design in common use. Prescriptive design is a "known." It is what has worked in the past. It matches other designs at existing facilities. AHJs are comfortable with prescriptive design and readily accept it.

One disadvantage of prescriptive design is that the safety factors can be so high as to render the design unduly expensive. A second disadvantage is that a prescriptive design might not result in the most effective way of protecting a particular facility. It neither accommodates a facility's specialized needs nor coordinates with other systems in the facility. The fire protection engineer's struggle with the efficacy of prescriptive design has helped support the trend toward performance-based design. (See Chapter 3.)

Prescriptive design is desirable so long as the advantages outweigh the disadvantages. For many facilities, prescriptive design can be fast and inexpensive. Its inherent safety factors can also provide sufficient flexibility for future changes. This type of design still serves light manufacturing facilities very well.

The more specialized the building, and the more its architecture departs from assumed norms, the higher the chance that performance-based design can better serve that building's fire protection needs.

4-2 Prescriptive Codes

Most prescriptive fire protection design is dictated through prescriptive codes. In the U.S., the prescriptive codes most often used in fire protection are the National Fire Protection Association (NFPA) codes and regional building codes. The regional building codes adopt many NFPA codes by reference. Some other countries also adopt NFPA codes, and some have their own comparable codes.

Prescriptive codes are both easy to apply and easy to misapply. The codes are straightforward, but the situations to which they apply might not be. In addition, several codes may apply simultaneously. Using some codes and leaving others out might compromise a design.

Probably the most familiar code that prescribes fire protection design is NFPA 13, *Standard for the Installation of Sprinkler Systems*. All major U.S. building and fire codes adopt NFPA 13 by reference. Just about everyone involved in building projects is familiar with this code.

Following the prescribed designs in NFPA 13 is an easy matter. However, using an NFPA 13 design does not necessarily ensure adequacy of fire protection. For one thing, the selected design must apply to the facility. For another, code provisions from sources other than NFPA 13 must also be met. (See Section 4-3, Design Coordination.)

Furthermore, NFPA 13 has more sprinkler design choices than ever before. Selecting the best one requires understanding all the options, their advantages and disadvantages, and their applications and limitations.

Some options have very limited applications, like ESFR technology and extended coverage sidewalls. Yet designers sometimes select these options whenever they appear to cost less, whether they apply to a project or not. This is unsatisfactory fire protection design.

Like any other code, NFPA 13 cannot be applied in isolation. For example, designing a sprinkler system based on a reported water flow and pressure is not sufficient. The fire protection engineer must also verify that the water supply is acceptable. For example, is the supply dedicated to fire protection or does it have mixed service use? Is it a gravity or pumping supply? How reliable is the water supply, and is the level of reliability appropriate for the facility?

Different problems can arise if the design specified in NFPA 13 cannot be met by the available water supply. Then a decision must be made whether to reinforce the supply, protect the facility by some other means, reduce the protection requirements of the facility by changing its design, or use a performance-based design. Options like these are best compared by the project design team early in the design.

Many other NFPA codes prescribe different facets of fire protection design. For example, NFPA 30, *Flammable and Combustible Liquids Code,*© covers storage and use of flammable liquids, and NFPA 2001 covers clean agent extinguishing systems. Some NFPA codes are occupancy-specific. For example, NFPA 318, *Standard for the Protection of Cleanrooms*, describes the protection required for cleanrooms. (See References.)

4-3 Inherent Risk

In contrast to using performance-based design, using prescriptive design does not require selecting an acceptable level of risk. For this and other reasons, many people believe that using prescriptive design totally eliminates any fire risk. This is not true. All prescriptive designs encompass an unstated, and usually uncertain, level of risk. All prescriptive codes encompass this risk within the code requirements.

Quantifying the risk in prescriptive designs is difficult, because applying the same fire protection recipe to different facilities results in as many levels of risk. Paradoxically, the risk inherent in prescriptive design can be estimated by using performance-based analysis. Having an idea of the level of risk involved in a prescriptive design is very important. For one thing, it allays the misperception of lack of risk. Secondly, it provides a base for valid comparison to performance-based alternatives that may be considered.

Prescriptive codes are still being written. Understanding the level of risk incorporated during code-writing could help make these codes more effective. For a discussion of the issue of risk in codes, see "The Importance of Risk Perceptions in Building and Fire Safety Codes," *Fire Protection Engineering* magazine, by Armin Wolski, Issue No. 10, Spring 2001.

Understanding the risk inherent in prescriptive design also paves the way for accepting performance-based design, where the level of risk is specified as a basis for the design. Future prescriptive fire protection codes and codes for all the other engineering disciplines will continue to merge prescriptive and performance-based design elements until the differences between the two types of design are imperceptible. This is why familiarity with prescriptive codes is not necessarily sufficient to adequately design engineered systems for a building.

The perceptions of the public have an effect on what codes require. This is why education of the public about risk has become increasingly important. See "Addressing Risk and Uncertainty in Performance-Based Fire Protection Engineering," *Fire Protection Engineering* magazine, by Brian Meacham, Ph.D., P.E., Issue No. 10, Spring 2001.

Othcr publications that address risk and uncertainty in fire protection include:

- *Introduction to Performance-Based Fire Safety*, Custer and Meacham
- NFPA *Fire Protection Handbook*
- SFPE *Handbook of Fire Protection Engineering*

4-4 Design Coordination

Because prescriptive fire protection codes can be followed like recipes, engineers and technicians in other disciplines are sometimes charged with applying them. However, because fire protection features interrelate with features of all the other building disciplines, applying prescriptive fire protection codes requires extensive coordination between the disciplines.

Traditionally, provisions for prescriptive fire protection design have been incorporated into codes that focus on the major discipline to which the provisions are most closely related. For example, electrical features of fire protection design are addressed by a group of codes that are electrically oriented. Chemical, structural, and mechanical features are likewise addressed by other groups of codes. This means that even simple prescriptive fire protection designs require knowledge of and coordination between many codes. (See Chapter 5.)

NFPA 13 is a good example of a prescriptive fire protection code requiring interdisciplinary coordination. Applying this code requires knowing the building construction and occupancy (fire loading). It also requires knowing enough about the hazards in the building to know when other codes apply.

Another example of the need for code coordination in fire protection design is in specifying a flammable liquids storage cabinet. The appropriate code, NFPA 30, *Flammable and Combustible Liquids Code,*© does not require the cabinet to be grounded. However, other codes may apply. If the cabinet is installed in a flammable liquids dispensing area, flammable vapors are likely to be present and NFPA 497, *Recommended Practice for the Classification of Flammable Liquids, Gases, or Vapors and of Hazardous (Classified) Locations for Electrical Installations in Chemical Process Areas*, would also apply. The cabinet would then have to be grounded.

Electrical engineers, who are usually very familiar with NFPA 70, *National Electrical Code,*© may not have a working knowledge of all the fire protection codes with provisions relating to electrical systems. Chapter 5 discusses many other areas where such coordination is required.

Several recipes are available for addressing life safety issues. Everyone working on a project must be aware of the recipe being used.

NFPA 101, *Life Safety Code,*© is referenced by the U.S. regional building codes. It is applied by fire protection engineers, as well as by architects and building designers. Although this code is often used, NFPA 101A, *Guide on Alternative Approaches to Life Safety*, is also available. This alternate code can be applied where accepted by the AHJ. NFPA 101A tailors life safety requirements to more specific conditions at a facility than NFPA 101, but it is still primarily a prescriptive code. The extent to which the alternative methods in NFPA 101A are used must be coordinated.

Chapter 9 of NFPA 101A presents a model called the Computerized Fire Safety Evaluation System (CFSES). This model is intended to compare the risks of various life safety design alternatives. It is one of just a few models that simulates flame spread. This is the start of performance-based life safety design alternatives to the more common prescriptive code.

NFPA 101B, *Means of Egress for Buildings and Structures*, is intended to be used with a building code that does not otherwise specify means of egress features.

The examples in this chapter give an idea of the extent of coordination required for the fire protection and fire protection-related aspects of a building project, even when using straightforward prescriptive codes. By training, fire protection engineers develop a familiarity with all the applicable codes and how they interface across the disciplines on a project. See Chapter 5 for many other features requiring coordination across the disciplines of a project.

5:
Interfacing With the Other Disciplines

5-1 Architectural

Project management for larger projects often rests with an architectural or A/E firm. The architect makes sure the buildings meet jurisdictional requirements as well as the functional needs of the owner. The architect must also make sure the project meets acceptable levels of risk in line with the owner's risk management goals. In doing this, architectural features must be coordinated with all the engineering designs.

Building architectural engineering is one of the newer engineering disciplines recognized by the National Council of Examiners for Engineering and Surveying (NCEES). The P.E. examination syllabus includes the following areas:

- General Knowledge of Building Systems, Materials, and Codes;
- Construction Management;
- Electrical Systems;
- Mechanical Systems; and
- Structural Systems.

The General Knowledge category includes an item for "fire protection systems relevant to electrical, mechanical, and structural design/components." This chapter discusses these systems.

Areas of concern to fire protection engineers that interface with architectural engineering include:

- Locations of buildings;
- Exposures to buildings;
- Sizes of buildings;
- Sizes of fire areas;

- Building layout;
- Combustibility of building finishing materials; and
- Security issues.

Architects make sure that projects meet the requirements of the appropriate building code. The building codes most frequently followed in the U.S. today are:

- *National Building Code*, published by BOCA, the Building Officials & Code Administrators International, Inc. This code is used mainly in the northeastern part of the U.S.
- *Standard Building Code*, previously called the Southern Standard Building Code, published by SBCCI, the Southern Building Code Congress International. This code is used mainly in the southeastern part of the U.S.
- *Uniform Building Code*, published by ICBO, the International Conference of Building Officials. This code is used in the western U.S.

Also note that many organizations, including the International Code Council (ICC), the Building Owners and Managers Association (BOMA), and the National Fire Protection Association (NFPA), support the development and use of one national building code in the U.S. The first edition of the *International Building Code* was published by the ICC in 2000. The first edition of *The NFPA 5000 Building Code* is currently out for review and comment.

Locations of Buildings

A fire protection concern in locating buildings is capability and response speed of fire fighting services. Two other concerns are proximity to fire protection water supplies and availability of fire alarm services.

Related concerns are the flood, earthquake, and wind zones in which the building falls. These are fire protection concerns because flood, earthquake, and wind incidents can start fires, make the results of fire worse, or affect the protection in service at a facility. Flood, earthquake and wind concerns are factors for other disciplines, as well.

Exposures to Buildings

Placement of buildings on a site dictates their exposure to and from other buildings, structures, and hazards. Examples of exposures that are not buildings or structures include yard storage, tank farms, and electrical substations.

Exposures can be from part of the facility being designed or modified, or they can be from another property. Exposures are usually considered to include those from wildfires and lightning. In many cases, exposures are also considered for flooding, earthquakes, and major windstorms. Effects of these exposures have a bearing on facility layout and design. (Also see Section 5-5.)

The following codes include guidance on protection from exposures.

- NFPA 80A, *Recommended Practice for Protection of Buildings from Exterior Fire Exposures*
- NFPA 299, *Standard for Protection of Life and Property from Wildfire*
- NFPA 780, *Standard for the Installation of Lightning Protection Systems*

Sizes of Buildings

The sizes of buildings can affect the risk from a single incident. The incident could be fire-related, business-related, or otherwise. While larger buildings are often better for process flow and building maintenance considerations, they can also increase the owner's risk. The risk must be weighed when making decisions on building size.

Sizes of Fire Areas

Sizes of fire areas present similar concerns to those on building size. Some fire separations are required by codes, some are required by AHJs, and some are desired by building owners for other reasons.

Building Layout

The interior building layout must meet applicable codes for life safety egress requirements. The usual code in the U.S. is NFPA 101, *Life Safety Code.*© NFPA 101A and/or NFPA 101B may apply instead, as may other codes. (See Chapter 4.)

Other factors can affect building layout. Walls may be used to restrict the extent of hazardous (classified) locations for the purpose of specifying electrical equipment. They may also be installed to restrict access to particular areas or to control how particular areas are accessed. In the end, both code and noncode provisions must be satisfied.

Hand in hand with the interior building layout comes locating fire extinguishers and hose standpipes. Such factors as size of areas, location of fire cutoffs, and occupancy hazards all affect the placement of these manual fire protection systems. The codes pertaining to these systems are:

- NFPA 10, *Standard for Portable Fire Extinguishers*
- NFPA 14, *Standard for the Installation of Standpipe, Private Hydrant, and Hose*

Combustibility of Finishing Materials

Selecting high flamespread interior finishing materials or insulation would negate the gains of specifying fire resistive or noncombustible building construction. The same codes that specify egress requirements also specify required characteristics of interior finishing materials for various types of buildings.

The following NFPA codes pertain to properties of interior finishing materials:

- NFPA 253, *Standard Method of Test for Critical Radiant Flux of Floor Covering Systems Using a Radiant Heat Energy Source*
- NFPA 255, *Test of Surface Burning Characteristics of Building Materials*
- NFPA 264, *Heat Release Rates for Materials*
- NFPA 265, *Evaluating Room Fire Growth Contribution from Textile Wall Coverings*
- NFPA 286, *Standard Methods of Fire Tests for Evaluating Contribution of Wall and Ceiling Interior Finish to Room Fire Growth*

Security Issues

Building design features for maintaining security affect fire protection in three primary ways:

- Means of egress for occupants;
- Access for firefighting; and
- Design of fire protection systems to eliminate their use as listening devices.

The first issue is addressed in NFPA 101, *Life Safety Code.*© Any security system provided must meet the applicable provisions of this code. NFPA 101 addresses such features as areas of refuge and delayed egress locks. Each occupancy chapter in NFPA 101 addresses security features that are acceptable for that occupancy. Many other NFPA codes address security issues.

Normally closed doors can add more confusion or delay in egress paths and may even block the view of potentially hazardous conditions on the other side of the door. Prisons or buildings with detention areas have many security concerns that must be balanced with fire protection and life safety needs.

Access for firefighting is addressed in:

- NFPA 1, *Fire Prevention Code*
- NFPA 601, *Standard for Security Services in Fire Loss Prevention*

Access to secure areas for firefighting purposes can be a very complex issue. If the security is to protect property or information, then the responding fire department can be given keys to the building or to key access boxes. Where it is not desirable for the fire department to keep doors open for bringing in hose lines, interior standpipes can be provided.

On the other hand, if the security is for protecting people against physical, radiological, or biological hazards, then more coordination is needed in letting the fire department into the building. People knowledgeable about the hazards must advise firefighters whether areas can be safely entered. For high hazard areas where firefighting is not likely to be safe, the building and its fire protection systems would have to be designed to function without human intervention.

In buildings with security boundaries all building systems, including fire protection systems, must be designed so that no part of the system can be used as a listening device. This includes piping, wiring, and ductwork, as well as radio transmitters, which can be found on alarm systems and control panels.

The book *Fire Safety and Loss Prevention*, by Kevin Cassidy, exclusively addresses the fire protection/security interface. (See References.)

Security measures can also benefit fire protection. Well secured areas should have fewer uncontrolled ignition sources, particularly from personnel. All facility personnel should be familiar with facility security measures and should be trained in how security affects operations, including emergency response.

5-2 Chemical

Fire is a chemical reaction, so it stands to reason that fire protection engineering has a large overlap with chemical engineering. Rare is the chemical process that does not introduce some type of hazard. Areas of concern to fire protection engineers that interface with chemical engineering include:

- Materials hazards;
- Materials storage;
- Process hazards evaluation; and
- Process utilities.

NFPA 45, *Standard on Fire Protection for Laboratories Using Chemicals*, addresses chemical hazards in the laboratory environment for all the above areas. The *NFPA Fire Protection Handbook* and other occupancy-related NFPA codes also address these areas.

Materials Hazards

Different types of materials burn differently. This may be for either physical or chemical reasons, or both. Physical reasons include the amount of surface area on the material, the fineness of the material, and the material's arrangement, configuration, and packaging. Chemical reasons include whether the material is a solid, liquid, or gas, the heat content, the rate of heat release when the material is burning, and the material's volatility, reactivity, flammable limits (for liquids), and ignitable concentrations (for dusts).

Many handbooks give the general physical and chemical properties of materials, including the following:

- *Chemical Engineers' Handbook*, Perry
- *CRC Handbook of Chemistry and Physics*
- *Lange's Handbook of Chemistry*
- *NFPA Fire Protection Guide to Hazardous Materials*

Fire protection-related physical and chemical properties are also found in:

- Drysdale, *An Introduction to Fire Dynamics*
- *NFPA Fire Protection Handbook*
- *NFPA Fire Protection Guide to Hazardous Materials*
- *SFPE Handbook of Fire Protection Engineering*
- N. Irving Sax, *Dangerous Properties of Industrial Materials*

Ordinary combustibles include wood, cardboard, and paper. They burn, but not so severely as to require protection beyond conventional sprinkler systems at the ceiling. The prescriptive method of designing conventional sprinkler systems (described in NFPA 13) includes classification of areas as Light Hazard, Ordinary Hazard Group 1, Ordinary Hazard Group 2, Extra Hazard Group 1, and Extra Hazard Group 2.

Buildings containing small amounts of ordinary combustibles only would normally fall into one of the first three categories. Buildings containing larger amounts of ordinary combustibles would require different designs. NFPA 13 addresses these designs. The extra hazard categories are suitable for only small amounts of hazardous materials. They do not provide unlimited protection for hazardous materials of any type or amount.

Protection of rolled paper and baled cotton is handled separately. Their configuration increases the fire hazard beyond ordinary. Plastics have their own classification system of Groups A, B, and C. These classes depend on the burning characteristics of the plastics. Plastics in the highest hazard category, Group A, burn fiercely while melting into a flammable liquid that quickly spreads the fire. Rubber tires have their own hazard category and protection criteria. NFPA 13 addresses protection of all these commodities.

Another category of materials presenting a fire and explosion hazard is combustible dusts. NFPA 499, *Recommended Practice for the Classification of Combustible Dusts and of Hazardous (Classified) Locations for Electrical Installations in Chemical Process Areas*, divides combustible dusts into three categories as follows:

- Group E, Atmospheres containing combustible metal dusts, including aluminum, magnesium, and their commercial alloys, or other combustible dusts whose particle size, abrasiveness, and conductivity present similar hazards in the use of electrical equipment.
- Group F, Atmospheres containing combustible carbonaceous dusts that have more than 8 percent total entrapped volatiles (see ASTM D 3175 for coal and coke dusts) or that have been sensitized by other materials so that they present an explosion hazard. Coal, carbon black, charcoal, and coke dusts are examples of carbonaceous dusts.
- Group G, Atmospheres containing other combustible dusts, including flour, grain, wood flour, plastic, and chemicals.

Metals are not normally considered combustible. However, most metals can burn under the right circumstances, such as when they are divided into sufficiently small or thin pieces and exposed to sufficient heat, oxygen and moisture. The metals referred to as combustible are those most easily ignited. These metals include lithium, magnesium, potassium, sodium, titanium, and zirconium. Sodium, potassium, and lithium are particularly volatile, due to their chemical characteristics as alkali metals.

Though not as easy to ignite, steel, aluminum, and other metals also burn. Fine metal shavings burn more readily in the presence of hydrocarbon-based oils. For example, fires are common on aluminum rolling mills. The fires usually start in the combustible rolling fluids, but they also involve the aluminum. The *NFPA Industrial Fire Hazards Handbook* includes fire hazards information on metals processing.

A few metals are pyrophoric (self-igniting in the presence of oxygen). These metals must be stored and handled under very controlled conditions, including keeping them under inert atmospheres. Pyrophoric metals include metal alkyls, alkali metals, and metal hydrides.

Relevant NFPA Codes for protecting combustible metals include:

- NFPA 480, *Storage, Handling, and Processing of Magnesium*
- NFPA 481, *Storage, Handling, and Processing of Titanium*
- NFPA 482, *Storage, Handling, and Processing of Zirconium*
- NFPA 485, *Standard for the Storage, Handling, Processing, and Use of Lithium Metal*
- NFPA 651, *Standard for the Machining and Finishing of Aluminum and the Production and Handling of Aluminum Powders*

The *NFPA Fire Protection Handbook* and the *NFPA Industrial Fire Hazards Handbook* also contain information on occupancies handling combustible metals.

Peroxides and other oxidizers increase the intensity of any fire they become involved in. This includes the gaseous oxygen found in the Earth's atmosphere, which is a particular hazard at higher than normal concentrations. Because of their very high reactivity, peroxides are used in making explosives.

Protection of oxidizers, peroxides and explosives is addressed in the following codes:

- NFPA 50, *Bulk Oxygen Systems at Consumer Sites*
- NFPA 53, *Fire Hazards in Oxygen-Enriched Atmospheres*
- NFPA 68, *Guide for Venting of Deflagrations*
- NFPA 69, *Explosion Prevention Systems*
- NFPA 99B, *Standard for Hypobaric Facilities*
- NFPA 430, *Storage of Liquid and Solid Oxidizers*
- NFPA 432, *Code for the Storage of Organic Peroxide Formulations*
- NFPA 490, *Storage of Ammonium Nitrate*

- NFPA 495, *Explosive Materials Code*
- NFPA 1124, *Code for the Manufacture, Transportation, and Storage of Fireworks*

Solids must usually be heated to emit vapors, except for combustible metals where chemical reactions release vapors. Most liquids readily emit vapors at ambient temperatures. Controlling the fire hazards of flammable and combustible liquids therefore involves controlling both the liquid and vapor phases.

Liquids themselves don't burn, their vapors do. The degree of hazard of a liquid depends on many properties, including its boiling point, flash point, autoignition temperature, and lower and upper flammable limits.

NFPA 497, *Recommended Practice for the Classification of Flammable Liquids, Gases, or Vapors and of Hazardous (Classified) Locations for Electrical Installations in Chemical Process Areas*, divides flammable gases and vapors into four groups. Note that these groups have no relation to the groups for classifying plastics.

- Group A: Acetylene.
- Group B: Flammable gas, flammable liquid-produced vapor, or combustible liquid-produced vapor mixed with air that may burn or explode, having either a maximum experimental safe gap (MESG) value less than or equal to 0.45 mm or a minimum igniting current ratio (MIC ratio) less than or equal to 0.40. (A typical Group B substance is hydrogen.)
- Group C: Flammable gas, flammable liquid produced vapor, or combustible liquid produced vapor mixed with air that may burn or explode, having either a maximum experimental safe gap (MESG) value greater than 0.45 mm and less than or equal to 0.75 mm, or a minimum igniting current ratio (MIC ratio) greater than 0.40 and less than or equal to 0.80. (A typical Group C substance is ethylene.)
- Group D: Flammable gas, flammable liquid produced vapor, or combustible liquid produced vapor mixed with air that may burn or

explode, having either a maximum experimental safe gap (MESG) value greater than 0.75 mm or a minimum igniting current ratio (MIC ratio) greater than 0.80. (A typical Group D substance is acetone.)

Flammable and combustible liquids are further divided into classes as follows:

- Class IA: Liquids having flash points below 73°F and boiling points below 100°F.
- Class IB: Liquids having flash points below 73°F and boiling points at or above 100°F.
- Class IC: Liquids having flash points at or above 73°F but below 100°F.
- Class II: Liquids having flash points at or above 100°F but below 140°F.
- Class IIIA: Liquids having flash points at or above 140°F but below 200°F.
- Class IIIB: Liquids having flash points at or above 200°F.

NFPA 30, *Flammable and Combustible Liquids Code*© contains the primary protection measures for combustible and flammable liquids. Other NFPA codes address specific operations involving flammable and combustible liquids.

Applicable codes for storing and using flammable and combustible liquids include the following:

- NFPA 30, *Flammable and Combustible Liquids Code*©
- NFPA 33, *Spray Application of Flammable and Combustible Materials*
- NFPA 34, *Dipping and Coating Processes Using Flammable or Combustible Liquids*
- NFPA 35, *Standard for the Manufacture of Organic Coatings*
- NFPA 77, *Static Electricity*
- NFPA 86, *Ovens and Furnaces*

- NFPA 96, *Installation of Equipment for the Removal of Smoke and Grease Laden Vapors from Commercial Cooking Equipment*
- NFPA 395, *Storage of Flammable and Combustible Liquids on Farms and Isolated Construction Projects*
- NFPA 496, *Purged and Pressurized Enclosures for Electrical Equipment in Hazardous Locations*
- NFPA 497, *Recommended Practice for the Classification of Flammable Liquids, Gases, or Vapors and of Hazardous (Classified) Locations for Electrical Installations in Chemical Process Areas*

Liquefied and compressed flammable gases present their own fire and explosion hazards. Codes relevant to protecting these hazards include:

- NFPA 50A, *Gaseous Hydrogen Systems at Consumer Sites*
- NFPA 50B, *Liquefied Hydrogen Systems at Consumer Sites*
- NFPA 51, *Design and Installation of Oxygen-Fuel Gas Systems for Welding, Cutting, and Allied Processes*
- NFPA 54, *National Fuel Gas Code*©
- NFPA 55, *Compressed and Liquefied Gases in Portable Cylinders*
- NFPA 58, *Storage and Handling of Liquefied Petroleum Gases*
- NFPA 59, *Storage and Handling of Liquefied Petroleum Gases at Utility Gas Plants*
- NFPA 59A, *Production, Storage and Handling of Liquefied Natural Gas*
- NFPA 86C, *Industrial Furnaces Using a Special Processing Atmosphere*

Pyrophoric gases are more hazardous than flammable gases because they are both flammable and self-igniting. The semiconductor industry uses many of these gases for doping semiconductor chips. The properties that make particular gases good semiconductor dopants also seem to make them pyrophoric.

The most common pyrophoric gases are the silanes, which are hydrides of silicon. NFPA 318, *Protection of Cleanrooms*, addresses the protection measures needed when using these gases. NFPA 55, *Compressed and Liquefied Gases in Portable Cylinders*, addresses storing these gases.

Another special material classification is for aerosols. The liquid aerosol product can be an inert, combustible or flammable liquid. The propellant can be an inert or flammable gas. Flammable aerosols propelled by flammable propellants are the most hazardous, but all aerosols present hazards. This is because when exposed to heat the containers rocket, presenting a projectile hazard and potentially spreading fire. NFPA 30B, *Aerosol Products, Manufacture and Storage*, addresses these hazards.

New materials are being developed with increasing frequency. It is almost impossible for general materials handbooks to stay up to date. Industries that develop and use new materials sometimes prepare their own handbooks. One example is the National Semiconductor *Chemical & Radiation Safety Handbook.* This handbook lists and describes many of the new chemicals being used in the semiconductor industry for manufacturing chips. Many of these chemicals are found in the general handbooks, but some of them are not.

Chemicals that do not present fire or explosion concerns may be toxic and/or carcinogenic. Like fire protection issues, these safety issues are of concern to all disciplines on a building project. Materials handbooks normally cover these properties in addition to the ones already mentioned.

A handbook dealing specifically with toxins and carcinogens is the NIOSH *Pocket Guide to Chemical Hazards*, published by the Center for Disease Control. NIOSH also publishes other chemical handbooks.

Materials Storage

Not only do physical and chemical characteristics affect the fire and explosion hazards of a material, but so does its configuration. Following are examples of configurations that affect the hazard of given materials:

- How the material is packaged, i.e., if it is in boxes with foam peanuts or cocoons, or whether it is shrink-wrapped;
- Whether the material is stored in bulk, on pallets, in double-row racks or in multiple-row racks;
- Aisle widths between piles or racks of storage;
- The dimensions of racks and their longitudinal and transverse flue spaces;
- Whether sprinkler coverage of material in racks is blocked by solid shelves, closed-top bins or other obstructions;
- How high the material is stored;
- How much material is in a given fire area;
- Whether flammable liquids are stored in small containers, drums, tanks, or large tank farms;
- Whether flammable liquids are stored in safety cans, safety cabinets, or properly protected flammable liquids rooms; and
- What else in a facility the storage exposes.

Process Hazards Evaluation

Processes involving chemicals present many hazards. These include the generation of unexpected heat or pressure, rupture of pipes and vessels, and many other problems. The *Kirk-Othmer Encyclopedia of Chemical Technology* is one of the most comprehensive guides on what chemicals are used in today's industrial processes and how they present hazards.

The most dangerous chemical reactions are exothermic, or heat-producing. The temperatures and pressures generated by these reactions must be controlled by proper introduction of raw material and by sufficient agitation and cooling. These processes must also be designed to withstand reasonable upsets. This is accomplished by reactor construction safety factors and by providing adequate normal and emergency vessel venting. One common resource for sizing reactor vents is the manual from the *Design Institute for Emergency Relief Systems* (DIERS).

Designing safe chemical processes involves considering many features, including the following:

- Hazards of raw materials, intermediates, and finished products;
- Hazards of material interactions;
- Designing for a minimum volume of hazardous material holdup;
- Safe temperature ranges and safety factors;
- Safe pressures and design of normal pressure/vacuum venting and emergency venting;
- Liquid flow rates and levels;
- Leak control through dikes, drains, and holding tanks;
- Cooling water system and backup system;
- Agitation required;
- Safety ventilation;
- Required sensors for monitoring unsafe conditions and interlocks for shutting down;
- Making hazardous features of the process failsafe; and
- Providing safe distances between hazards and exposed property.

The safety of a process is not assured until the management programs are in place to support it. These programs include:

- Normal startup, operating, and shutdown procedures, including safety checks;
- Safe off-normal and emergency procedures;
- Operator training and refresher training in the procedures;
- Inspection, testing, and preventive maintenance of process and safety equipment; and
- Management of change.

Process safety features are best considered at the time of facility and/or process design to enable specifying the correct equipment and designing a proper layout. They apply to simple processes as much as they apply to the more complex ones.

Taking the last item, management of change to a process involves analyzing all possible effects each change might have on that process. This amounts to conducting a new process hazards evaluation. Common examples of process changes that warrant a process hazards evaluation are:

- Using different raw materials, solvents, catalysts, or other substances;
- Changing how materials flow through the process;
- Operating at different speeds, temperatures, or pressures;
- Increasing or decreasing equipment tolerances;
- Adding new subsystems, such as coating, drying, or drainage systems;
- Changing the characteristics of the finished product;
- Revising maintenance schedules, frequencies, or methods; and
- Changing the area ventilation, construction, or other characteristics.

Considering these features before selecting equipment can result in better design choices.

When the project involves existing equipment, a common example that might not be flagged for a new process hazards evaluation is replacement of a part with a like part. However, this too warrants a new process hazards evaluation for the following reasons:

- The process may have been adjusted or reworked due to wear in the part being replaced.
- The replacement part may not be identical in all respects to the part it is replacing. Manufacturers of parts frequently change manufacturing methods, gasket materials or other characteristics

of parts. They could also change a part's pressure rating or their specifications for how a part should be installed or calibrated.

- The people installing the part may not have been present at the initial process setup. Their review of a new process hazards evaluation will familiarize them with loss scenarios involving that part, as well as the rest of the process.
- A better part may have been developed and may have become available since the process was first set up. This option should always be checked before automatically assuming a straight replacement should be done.
- The need to replace a part could be an opportunity for improving other process features, sometimes at reduced cost. This possibility should always be reviewed.

Process Utilities

Chemical processes often need specialized utilities, which can present their own fire and explosion hazards. Some processes may use conventional fuels, others may use special fuels or fuel mixes. Some process raw materials are purchased in bulk from utility suppliers. This is fairly common for liquid oxygen, liquid nitrogen, inert gases, and semiconductor gases. Liquids can also be purchased this way.

Many chemical processes require close temperature control. Facilities can buy or make their own steam, or they can use heat transfer fluids. The most common heat transfer fluids are combustible oils, which can present serious problems if they are overheated or if they leak from piping or containment vessels. Other processes use combustible hydraulic fluids, which are usually under very high pressure.

Codes pertaining to heat transfer systems include:

- NFPA 30, *Flammable and Combustible Liquids Code*©
- NFPA 61, *Standard for the Prevention of Fires and Dust Explosions in Agricultural and Food Products Facilities*

- NFPA 70, *National Electrical Code©*
- NFPA 86, *Standard for Ovens and Furnaces*
- NFPA 86C, *Standard for Industrial Furnaces Using a Special Processing Atmosphere*
- NFPA 86D, *Standard for Industrial Furnaces Using Vacuum as an Atmosphere*
- NFPA 664, *Standard for the Prevention of Fires and Explosions in Wood Processing and Woodworking Facilities*

Note that the occupancy-related codes for agricultural and wood processing facilities address fire prevention design features for heat transfer systems that could also apply to other types of facilities.

The cost of discharging hazardous waste encourages treatment or recovery of process materials like catalysts and solvents. Common recovery measures include filtering, collection, and distillation systems. These operations can be very hazardous and must be protected accordingly.

Finally, chemical processes often depend on pollution control measures like incinerators, flare stacks, precipitators, and other pollution control equipment. Protection for these potentially hazardous operations downstream of the actual process must be included in the process design.

The process hazards analysis must consider all process utility systems. Controlling the hazards from utility systems requires appropriate design; adequate inspection, testing, and maintenance procedures; and good operator training programs. In addition, utility systems must be reviewed from a potential business interruption standpoint. For example, even if a particular problem with a utility system is not expected to cause a fire, it could have the potential to shut a process down for an unacceptably long period of time.

5-3 Electrical

The bulk of the electrical work for most building projects is usually designing ordinary circuits for building electrical power and laying out the lighting systems. Electrical systems for some types of buildings, like hospitals, also have special requirements that engineers must learn about. For heavy industry, electrical design could also include designing large power distribution systems for feeding power-intensive process equipment.

Electrical engineers are very familiar with NFPA 70, *National Electrical Code*© (NEC), codes published by the International Electrotechnical Commission (IEC) or comparable electrical codes (for other countries outside the U.S. and Europe). The requirements in the applicable generic code are sufficient for the design of most electrical systems.

Additional requirements apply to wires and cables used in air-handling spaces. The NFPA code addressing this subject is NFPA 262, *Standard Method of Test for Flame Travel and Smoke of Wires and Cables for Use in Air-Handling Spaces.*

Because fire alarm, security, and fire protection control systems must be wired, their design often falls to electrical engineers. Meeting the generic electrical code is not sufficient. To design these systems properly, the electrical engineers must be familiar with a host of electrically related fire protection requirements.

Fire protection engineering interfaces with electrical engineering in the following areas:

- Building system controls;
- Fire detection and alarm systems;
- Extinguishing system control;
- Electric motor-driven fire pumps;
- Emergency lighting;
- Backup power supplies;

- Electrical equipment for hazardous (classified) locations; and
- Electrical protection systems.

Building System Controls

Controls for various building systems present both electrical and fire protection concerns. These systems include the following:

- HVAC systems;
- Smoke control systems;
- Cranes; and
- Elevators.

HVAC systems are now often arranged to perform energy management functions. Smoke control can be handled by HVAC systems or by separate systems. The electrical portions of HVAC and smoke management systems are important because the controls dictate the action of the system.

Controls are also susceptible to being changed by people, which introduces the potential for eliminating intended design features or for creating new hazards. Because these are mechanical systems, they are also discussed in Section 5-4.

Operations in many facilities require permanently installed cranes. The cranes can be inside, outside, or both. Crane controls are very important for prevention of many types of losses, including fires. Crane wiring must often be exposed, so it must be well grounded and protected. The logic of the electrical control system protects the crane and the building, as well as the building contents. Operator training is also very important.

Control systems for elevators must be in compliance with NFPA 101, *Life Safety Code.*© This ensures the safety of building occupants as well as responding firefighters.

Fire Detection and Alarm Systems

The applicable code for design of detection and alarm systems in the U.S. (and in some other countries) is NFPA 72, *National Fire Alarm Code.*© For detection systems, this code contains provisions for:

- Selection of detectors;
- Location and spacing of detectors;
- Wiring of the detector circuits;
- Wiring to the detection control panel;
- Back-up power requirements;
- Arrangement, location, and wiring of audible and visible alarms; and
- Location of manual devices.

More types of fire detectors are available than ever before. Selecting the most appropriate detector for the application requires extensive knowledge of detector types, how they work, and their advantages and disadvantages.

For alarm systems, NFPA 72 contains provisions for:

- Arranging initiating device and signaling line circuits;
- Monitoring circuit integrity;
- Back-up power requirements; and
- Requirements related to the type of alarm service, e.g., central station, remote supervising station, proprietary supervising station, or other type of service.

In all cases, equipment should be listed for its intended purpose. For example, alarm system control panels can be listed for central station, remote supervising station, or proprietary supervising station service, or for any combination of these services. A central station control panel should have a listing for central station use.

Writing proper electrical specifications requires confirming the type of alarm system to be provided and understanding the relevant code requirements. Many other requirements can pertain to alarm systems, including what conditions the alarm system should monitor and whether the system is to be certificated or placarded. Specifications should usually include listing by nationally recognized testing laboratories, the most common of which is Underwriters' Laboratories.

A potential source of confusion in listings for fire alarm equipment is that many different types of listings apply. Separate listings are provided for protected premises fire alarm systems and for three types of supervising station fire alarm systems. Other listings are also provided for public reporting systems, security systems, and burglar alarm systems. Specifying a listing is not sufficient; the type of listing must also be specified.

Extinguishing System Control

Fire protection special extinguishing systems include those using alternate (gaseous) agents, carbon dioxide, dry chemical, foam, foam-water, water, and wet chemical. The NFPA codes for these types of systems include:

- NFPA 11, *Standard for Low Expansion Foam*
- NFPA 11A, *Standard for Medium- and High-Expansion Foam Systems*
- NFPA 12, *Standard on Carbon Dioxide Extinguishing Systems*
- NFPA 12A, *Standard on Halon 1301 Fire Extinguishing Systems*
- NFPA 15, *Standard for Water Spray Fixed Systems for Fire Protection*
- NFPA 16, *Standard for the Installation of Foam-Water Sprinkler and Foam-Water Spray Systems*
- NFPA 17, *Standard for Dry Chemical Extinguishing Systems*
- NFPA 17A, *Standard for Wet Chemical Extinguishing Systems*
- NFPA 750, *Standard on Water Mist Fire Protection Systems*
- NFPA 2001, *Standard on Clean Agent Fire Extinguishing Systems*

The electrical control of these systems is covered in the above codes. Factors to consider when designing these control systems include the factors listed in the section on fire detection and alarm systems, plus the following additional features:

- Control panel listing for release device service and for use with the type of release mechanism on the selected extinguishing system;
- Special arrangement of detection circuits, e.g., cross-zoning or other means of confirming an alarm signal;
- Presence of time delays, aborts, or maintenance disable switches;
- System interlocks to be provided, e.g., door and damper closure, electrical power shut-off, HVAC shutdown; and
- Monitoring by the building alarm panel.

Cross-zoning is now a generic term for any algorithm that requires two detectors to operate before an extinguishing agent is discharged. These control panel algorithms were designed to reduce chances of unnecessary discharge of extinguishing agent. Such algorithms are only appropriate for conventional spot-type smoke detectors. They should not be used for beam-type or high sensitivity smoke detectors, or for heat, flame, or spark detectors.

Including time delays, aborts, and maintenance disable switches in an extinguishing system can be very controversial. Designing them becomes particularly complex when two detectors must operate to discharge the extinguishing agent. The designer must coordinate design decisions about these features with the owner and the owner's authority having jurisdiction.

Electric Motor-Driven Fire Pumps

NFPA 20, *Standard for the Installation of Stationary Pumps for Fire Protection*, contains special requirements for wiring power feeds to electric fire pump controllers. The requirements include how the circuits should be arranged (wired), controlled (switched), and protected against exposure to damage.

Issues of critical importance concerning electric motor-driven fire pumps are:

- Connecting the pump's power feed upstream of manual shut-off switches, so that the pump runs even if all remaining power has been shut off;
- Setting circuit breakers so that the pump motor's inrush current does not trip them;
- Setting circuit breakers so that the circuit remains energized for a specified duration at locked rotor current; and
- Properly arranging and installing automatic transfer switches.

Many fire pump controllers have settings that need to be specified. Decisions about how to set fire pump controllers should be made in consultation with the owner and the authority having jurisdiction.

Emergency Lighting

NFPA 101, *Life Safety Code*© requires emergency lighting so that even in the event of loss of electrical power to a facility, occupants have enough light to safely exit the building. The code offers several options for meeting backup power requirements.

Backup Power Supplies

Some occupancies require backup power supplies for critical systems. For example, NFPA 99, *Standard for Health Care Facilities*, requires backup power in certain health care occupancies. The following codes cover backup power supplies:

- NFPA 110, *Standard for Emergency and Standby Power Systems*
- NFPA 111, *Standard on Stored Electrical Energy Emergency and Standby Power*

Even when facilities do not have full backup power supplies, individual systems are required to have their own backup power supplies, including fire protection systems. Emergency lighting systems are a good example. It is also common for the AHJ to require backup power supplies for other operations, such as ventilation systems, smoke control systems or critical process safety controls.

In addition, many other fire protection systems require backup power supplies. Examples of such systems include fire alarm systems and special extinguishing control systems. The specific requirements for these backup systems are found in the codes pertaining to these systems. For the alarm system example, the pertinent code would be NFPA 72, *National Fire Alarm Code.©* Extinguishing systems are addressed in several NFPA codes, including the following:

- NFPA 11, *Standard for Low Expansion Foam*
- NFPA 11A, *Standard for Medium- and High-Expansion Foam Systems*
- NFPA 12, *Standard on Carbon Dioxide Extinguishing Systems*
- NFPA 12A, *Standard on Halon 1301 Fire Extinguishing Systems*
- NFPA 15, *Standard for Water Spray Fixed Systems for Fire Protection*
- NFPA 16, *Standard for the Installation of Foam-Water Sprinkler and Foam-Water Spray Systems*
- NFPA 17, *Standard for Dry Chemical Extinguishing Systems*
- NFPA 17A, *Standard for Wet Chemical Extinguishing Systems*
- NFPA 750, *Standard on Water Mist Fire Protection Systems*
- NFPA 2001, *Standard on Clean Agent Fire Extinguishing Systems*

Electrical Equipment in Hazardous (Classified) Locations

For the purpose of specifying electrical equipment, hazardous (classified) locations are locations likely to contain flammable vapors, ignitable dusts, or combustible fibers and flyings. The National Electrical Code specifies types of electrical equipment suitable for these locations. However, it does not specify how to determine the extent of classified locations.

Guidance on classifying locations on and selecting appropriate electrical equipment is given in:

- NFPA 497, *Recommended Practice for the Classification of Flammable Liquids, Gases, or Vapors and of Hazardous (Classified) Locations for Electrical Installations in Chemical Process Areas*
- NFPA 499, *Recommended Practice for the Classification of Combustible Dusts and of Hazardous (Classified) Locations for Electrical Installations in Chemical Process Areas*

Hazardous areas are classified by type of exposure:

- Class I for flammable liquid vapors;
- Class II for ignitable dusts; and
- Class III for combustible fibers and flyings.

Class I is divided into Groups A, B, C, and D depending on the gas or vapor's specific gravity and required ignition energy. NFPA 497 gives the classifications for most liquids. Class II is divided into Groups E, F, and G, depending on the characteristics of the dust. NFPA 499 gives the classification of most dusts.

Hazardous area classifications are further subdivided by likelihood of exposure to the hazardous material. Two systems of subdivision are in common use: the older NEC system and the newer IEC/NEC system. The first system has Divisions 1 and 2, the other Zones 0, 1, and 2. (Note that Division 1 and Zone 1 are not equivalent.) Both these systems are recognized in the latest NFPA 70, *National Electrical Code.*© Specifying the correct division requires knowing which system of classification is being used.

Many types of electrical equipment are suitable for hazardous (classified) locations. The most well-known is explosion-proof equipment, which is commonly used in areas likely to contain flammable liquids vapors. The analogous equipment used in areas likely to contain dusts is called dust-ignitionproof.

Another type of electrical equipment suitable for hazardous locations is purged or pressurized equipment. NFPA 496, *Standard for Purged and Pressurized Enclosures for Electrical Equipment*, contains the requirements for this type of equipment.

A potential area of confusion with respect to electrical listings is in whether a piece of equipment is listed for use in hazardous (classified) locations, or is listed for containing hazardous materials. One example is in the listing of refrigerators. Refrigerators listed under category STRV of the *UL Hazardous Locations Equipment Directory* are suitable for use in hazardous (classified) locations. Refrigerators listed under category SOVQ of the *UL Electrical Appliance and Utilization Equipment Directory* are suitable for containing flammable liquids.

A common area of misunderstanding in the electrical field is in what constitutes intrinsically safe electrical equipment. Most engineers know the basic requirement that energy from the equipment cannot ignite a hazardous atmosphere. But they may not know that the design depends on the division or zone, or how to determine the fault currents. Requirements for this equipment are specified in ANSI/UL 913. Nationally recognized testing laboratories list equipment that meets this standard as intrinsically safe. Ordinary low voltage equipment that is not so listed, including battery-operated flashlights, is not intrinsically safe.

Electrical Protection Systems

Electrical protection systems include systems for grounding, surge protection and lightning protection. These systems protect buildings and contents from electrical damage; they also reduce the potential for electrical ignition of a fire.

Grounding systems prevent accumulation of static charges that could ignite vapors. They can be designed to protect equipment and personnel, and they can ensure appropriate power quality. Grounding systems are an essential part of systems for surge protection and lightning protection. See NFPA 77, *Recommended Practice on Static Electricity*.

Surge protection is installed primarily to protect electrical equipment from damage due to unpredictable swings in the normal electrical power supply. It maintains acceptable power quality. This protection can also keep equipment from smoking, burning out, or starting a fire. Surge protection is not the same as lightning protection.

Buildings and other structures are often protected against lightning, particularly in areas prone to frequent lightning strikes or for structures likely to contain ignitable atmospheres. Traditional lightning protection systems use receptors (lightning rods), main conductors, and ground terminals to carry the strike current safely to earth.

Applicable references for lightning protection are:

- NFPA 780, *Standard for the Installation of Lightning Protection Systems*
- UL 96A, *Standard for Installation Requirements for Lightning Protection Systems*
- LPI-175, *Lightning Protection Institute Standard Practice*

Lightning is extremely unpredictable. Occasionally, even areas considered to be fully protected against lightning have suffered damage. Traditional lightning protection is acknowledged to be imperfect. Nontraditional methods for providing lightning protection include charge dissipation and early streamer emissions systems. These methods are controversial and have not yet been incorporated in traditional codes.

5-4 Mechanical

The most well-known overlap between the mechanical and fire protection disciplines is in laying out conventional sprinkler systems, but there are other areas of mutual concern. Fire protection systems that interface with the mechanical discipline include:

- Piped fire protection systems;
- Fire protection water supplies;
- Pneumatic power and control systems;
- Building HVAC systems;
- Smoke control and smoke management systems; and
- Area ventilation systems.

Piped Fire Protection Systems

Piped fire protection systems that are water-based include wet and dry pipe sprinkler systems, deluge and preaction sprinkler systems, and open- and closed-head water spray systems. Each type of system has a particular application.

The NFPA codes pertaining to water-based fire protection systems include:

- NFPA 11, *Standard for Low Expansion Foam*
- NFPA 11A, *Standard for Medium- and High-Expansion Foam Systems*
- NFPA 13, *Standard for the Installation of Sprinkler Systems*
- NFPA 15, *Standard for Water Spray Fixed Systems for Fire Protection*
- NFPA 16, *Standard for the Installation of Foam-Water Sprinkler and Foam-Water Spray Systems*
- NFPA 750, *Standard on Water Mist Fire Protection Systems*

Some publications categorize piped fire protection systems as either sprinkler systems or special extinguishing systems. The special extinguishing systems would include the foam and water spray systems bulleted above, as well as the nonwater-based systems. Piped fire protection systems that are nonwater-based include CO_2, Halon, dry and wet chemical, and clean agent systems.

The NFPA codes pertaining to the nonwater-based fire protection systems include:

- NFPA 12, *Standard on Carbon Dioxide Extinguishing Systems*
- NFPA 12A, *Standard on Halon 1301 Fire Extinguishing Systems*
- NFPA 17, *Standard for Dry Chemical Extinguishing Systems*
- NFPA 17A, *Standard for Wet Chemical Extinguishing Systems*
- NFPA 69, *Explosion Prevention Systems*
- NFPA 2001, *Standard on Clean Agent Fire Extinguishing Systems*

NFPA codes that cover features above and beyond the piped fire protection systems are also relevant. The most common example is NFPA 230, *Standard for the Fire Protection of Storage*. Other occupancy-related NFPA codes would also apply.

Relevant listings for fire protection piping, valves, and other devices are found in the *UL Fire Protection Equipment Directory*. The *FM Approval Guide* contains similar listings, as do directories published by other nationally recognized testing laboratories.

Fire Protection Water Supplies

Fire protection water supplies include water storage tanks, pumps, underground fire protection mains, hydrants, and feeds to sprinkler systems. NFPA codes pertaining to fire protection water supplies include:

- NFPA 13, *Standard for the Installation of Sprinkler Systems*
- NFPA 14, *Standard for the Installation of Standpipe, Private Hydrant, and Hose Systems*
- NFPA 20, *Standard for the Installation of Stationary Pumps for Fire Protection*
- NFPA 22, *Standard for Water Tanks for Private Fire Protection*

- NFPA 24, *Standard for the Installation of Private Fire Service Mains and Their Appurtenances*

Pneumatic Power and Control Systems

In hazardous (classified) areas where ignition sources must be eliminated, pneumatic systems are often used as power sources for opening and closing doors or for positioning machinery. Pneumatic systems can also be used for process control to eliminate the ignition sources presented by electrically operated control panels.

A common misconception is that all low voltage control panels are intrinsically safe. However, this is only true if the panel is designed to particular standards and is listed as intrinsically safe. (See the Electrical section of this chapter.) Finally, pneumatic systems can be used for conveying combustible or noncombustible particulates.

Although there is no NFPA code specifically for pneumatic systems, over 50 NFPA codes address them, including the several codes on life safety and health care facilities. Codes on hazardous occupancies, extinguishing systems, and alarm and detection systems also address pneumatic systems.

Two examples are:

- NFPA 45, *Standard on Fire Protection for Laboratories Using Chemicals*
- NFPA 650, *Standard for Pneumatic Conveying Systems for Handling Combustible Particulate Solids*

Building HVAC Systems

The design of some building HVAC systems must take fire protection features into account. One example is when the HVAC system serves hazardous (classified) areas. The parts of the HVAC system serving these areas must be properly isolated from those serving nonhazardous areas.

Sometimes hazardous areas have standalone HVAC units, which must be appropriate for the hazard. Another example is when areas of a facility are maintained at a higher pressure than the rest of the building for the purpose of keeping vapors or dusts from entering areas not designed for that hazard. Finally, HVAC systems can be arranged to shut down to prevent spreading of smoke.

Protection of the boilers that provide the source of building heat is addressed in NFPA 85, *Boiler and Combustion Systems Hazards Code*.

Building HVAC systems are often adapted to provide some smoke control functions. This is in contrast to dedicated smoke control systems, which are discussed in the next section.

Requirements for different types of building HVAC systems are covered in the following NFPA codes:

- NFPA 90A, *Standard for the Installation of Air-Conditioning and Ventilating Systems*
- NFPA 90B, *Standard for the Installation of Warm Air Heating and Air-Conditioning Systems*
- NFPA 91, *Standard for Exhaust Systems for Air Conveying of Vapors, Gases, Mists, and Noncombustible Particulate Solids*
- NFPA 96, *Ventilation Control and Fire Protection of Commercial Cooking Operations*
- NFPA 318, *Standard for the Protection of Cleanrooms*

Smoke Control and Smoke Management Systems

NFPA 92A, *Recommended Practice for Smoke-Control Systems*, addresses smoke control by using barriers, air flows, and pressure differences to confine smoke to the zone of fire origin and thus maintain a tenable environment in other zones. Smoke control systems require specifically designed ducts and/or plenums, control systems, and control system logic.

NFPA 92B, *Guide for Smoke Management Systems in Malls, Atria, and Large Areas*, addresses smoke management in large-volume, noncompartmented spaces. Rather than focusing on controlling the path of any smoke, this standard explains how to determine where smoke will go for a given building design. The idea is to design buildings where smoke accumulates outside of egress paths. The primary application of this standard is for stadia, malls, and atria.

NFPA codes that address generation and venting of smoke include:

- NFPA 204, *Standard for Smoke and Heat Venting*
- NFPA 262, *Standard Method of Test for Flame Travel and Smoke of Wires and Cables for Use in Air-Handling Spaces*
- NFPA 271, *Standard Method of Test for Heat and Visible Smoke Release Rates for Materials and Products Using an Oxygen Consumption Calorimeter*

Area Ventilation Systems

Mechanical ventilation is often needed for protection from fire and explosion in localized areas. A classic example is for the ventilation systems specified in NFPA 30, *Flammable and Combustible Liquids Code,*© for flammable liquids dispensing rooms. Another common example is the safety ventilation required by NFPA 86, *Standard for Ovens and Furnaces*, for ovens drying off flammable vapors.

Other applications for area ventilation are around process equipment that could release toxic or carcinogenic vapors, or around hazardous gas cylinders stored inside a facility. NFPA codes relevant to the occupancy address specific requirements.

Ventilation is required for rooms housing battery banks, because of the possibility that the batteries generate hydrogen. The following codes cover the ventilation requirements for backup power supplies:

- NFPA 110, *Standard for Emergency and Standby Power Systems*
- NFPA 111, *Standard on Stored Electrical Energy Emergency and Standby Power*

Sometimes specialized ventilation is used to reduce the chance of vapors or dusts of approaching hazardous concentrations near a process area. This allows for use of less stringently rated electrical equipment in that area, as well as other potential sources of ignition. Such ventilation systems must be interlocked so that the process capable of generating vapors or dusts is shut down if the ventilation system is not operating. Dedicated ventilation systems are also provided for exhaust systems on paint spray booths, over dip tanks, and on laboratory benches using hazardous materials.

Two of many NFPA codes addressing bench exhaust systems are:

- NFPA 45, *Standard on Fire Protection for Laboratories Using Chemicals*
- NFPA 318, *Protection of Cleanrooms*

Another application of mechanical systems for fire safety is in deflagration venting of process vessels. In this case, the venting system must be designed to withstand the maximum expected pressure developed in a deflagration. NFPA 68, *Guide for Venting of Deflagrations*, covers these systems.

Buildings or rooms can also be designed for deflagration venting. It should be noted that no type of design for deflagration works if the material presenting the hazard can detonate. See *Deflagration Venting for Buildings* in the next section.

5-5 Structural

Relevant fire protection questions about a structure are: Will the structure itself add to the facility's combustible loading? If so, how much? If not, how much heat can the structure withstand? If a fire started, how far would it be possible to spread? The interface between fire protection and structural engineering includes the following specific areas:

- Combustibility/fire resistance of building structural materials;
- Fireproofing of building structural elements;
- Fire resistance ratings of barriers;
- Protection of openings;
- Deflagration venting for buildings; and
- Flood, earthquake, snow, and wind design.

Combustibility/Fire Resistance of Building Structural Materials

No building material meets all possible needs. Pound for pound, wood is superior for withstanding earthquakes and other structural stresses, but it is combustible. Lightweight noncombustible materials fail quickly when exposed to heat. Heavy noncombustible materials are expensive and can be labor intensive to install. Building construction materials must be chosen to meet the applicable codes, local conditions, architectural needs, and the owner's goals.

NFPA codes relevant to structural features include:

- NFPA 203, *Guide on Roof Coverings and Roof Deck Construction*
- NFPA 220, *Types of Building Construction*
- NFPA 251, *Fire Tests of Building Construction Materials*
- NFPA 256, *Methods of Fire Tests of Roof Coverings*
- NFPA 259, *Test Method for Potential Heat of Building Materials*
- NFPA 268, *Standard Test Method for Determining Ignitability of Exterior Wall Assemblies Using a Radiant Heat Energy Source*

Underwriters Laboratories lists noncombustible roofing materials and systems in the *UL Roofing Materials and Systems Directory*. UL Fire Classified roofs are rated for internal fire exposure. UL Class A, B, or C roof coverings are rated for external fire exposure. The *FM Approval Guide* designation of Class 1 applies to both internal and external fire exposure.

Fireproofing of Building Structural Elements

If a selected building construction material does not provide the fire resistance desired for a particular part of a building, fireproofing can be used in that part. Information on fireproofing building structural elements is found in the *Underwriters Laboratories Fire Resistance Directory*.

Fireproofing systems for withstanding fires in ordinary combustibles are tested in accordance with ANSI/UL Standard 263 and listed in the directory under category BXUV. Each fireproofing system is designated by a design letter followed by three numbers. The letter specifies the construction element (walls, floors, ceilings, beams, columns) and the number specifies the type of fireproofing (membrane protection, direct applied, unprotected).

Fireproofing systems for withstanding rapid temperature rise fires are tested in accordance with ANSI/UL 1709 and listed in the directory under category BYBU. The system designation is the same as for ANSI/UL 263 systems except that the letter R is added after the first letter. The engineer must be sure to specify the correct fire exposure (ordinary or rapid temperature rise) for the building of concern. As an alternate to using listed systems, fire resistance ratings can be calculated from first principles. However, the correct fire exposure must still be used.

The *FM Approval Guide* contains information similar to that in the UL directories, as do the listings published by other nationally recognized testing laboratories.

Fire Resistance Ratings of Barriers

Fire resistive walls or barrier walls are sometimes used to divide buildings into separate fire areas. Fire walls are used for various reasons, including:

- Dividing spaces into areas for different tenants;
- Separating egress paths from other areas of a building;
- Separating office, manufacturing, and storage areas;
- Dividing manufacturing buildings into areas of different occupancies or processes;
- Separating service areas (boiler rooms, transformer rooms) from the remainder of a facility;
- Separating high hazard occupancies (flammable liquids vaults) from the remainder of a facility; or
- Subdividing a large area of similar occupancy to limit possible fire loss.

NFPA codes pertaining to fire walls and fire barrier walls are:

- NFPA 221, *Fire Walls and Fire Barrier Walls*
- NFPA 251, *Fire Tests of Building Construction Materials*

Also see the *Underwriters Laboratories Fire Resistance Directory*, *FM Approval Guide*, and listings published by other nationally recognized testing laboratories.

Protection of Openings

Moving people and materials through a facility requires that interior walls have openings. Routing utilities through a facility sometimes requires wall penetrations. Openings and penetrations in fire barrier walls need protection that provides fire resistance comparable to that of the wall.

Many types of protection for openings are available, including the following:

- Fire doors;

- Assemblies with fire-rated glass;
- Fire shutters for conveyor openings and other small openings;
- Fire dampers for ducts; and
- Fire-rated materials for sealing duct, pipe, and wire penetrations.

Note that in the strict sense of the term, a true fire wall (as opposed to a fire barrier wall) must be self-supporting and can have no penetrations. This is so that a building collapse on one side of the wall does not affect the other side. True fire walls can have protected openings, but they should be kept to a minimum.

The following NFPA codes address opening protection:

- NFPA 80, *Standard for Fire Doors and Fire Windows*
- NFPA 252, *Fire Tests of Door Assemblies*
- NFPA 257, *Fire Tests of Window Assemblies*

The *Underwriters Laboratories Building Materials Directory* lists fire doors under category GSNV and fire dampers under category EMME. The *Underwriters Laboratories Fire Resistance Directory* lists fire-rated penetration protection under Through-Penetration Firestop Systems (XHEZ), Fill Void or Cavity Materials (XHHW), and Firestop Devices (XHJI). UL also lists Perimeter Fire Containment Systems (XHDG). The *FM Approval Guide* lists similar systems for opening protection, as do other nationally recognized testing laboratories.

Deflagration Venting for Buildings

Deflagration venting limits building structural damage by allowing an intentionally weak part of a building to blow out while leaving the rest of the structure intact. Deflagration venting is often provided for flammable liquids storage or dispensing rooms or for buildings containing hazardous processes.

Deflagration venting is often called explosion venting, but that is a misnomer. In deflagrations, propagation of the combustion zone is less than the speed of sound. In detonations, propagation of the combustion zone is greater than the speed of sound. Deflagrations can be successfully vented in low strength enclosures, whereas detonations cannot be.

NFPA 68, *Guide for the Venting of Deflagrations*. NFPA 69, *Standard on Explosion Prevention Systems*, also discusses deflagration venting. It includes information on deflagration suppression systems and deflagration pressure containment.

NFPA codes that address when deflagration venting should be provided include, but are not limited to, the following:

- NFPA 15, *Standard for Water Spray Fixed Systems for Fire Protection*
- NFPA 30, *Flammable and Combustible Liquids Code*©
- NFPA 30B, *Code for the Manufacture and Storage of Aerosol Products*
- NFPA 33, *Standard for Spray Application Using Flammable or Combustible Materials*
- NFPA 35, *Standard for the Manufacture of Organic Coatings*
- NFPA 45, *Standard on Fire Protection for Laboratories Using Chemicals*
- NFPA 61, *Standard for the Prevention of Fire and Dust Explosions in Agricultural and Food Products Facilities*
- NFPA 86, *Standard for Ovens and Furnaces*
- NFPA 91, *Standard for Exhaust Systems for Air Conveying of Vapors, Gases, Mists, and Noncombustible Particulate Solids*
- NFPA 318, *Standard for the Protection of Cleanrooms*
- NFPA 654, *Standard for the Prevention of Fire and Dust Explosions from the Manufacturing, Processing, and Handling of Combustible Particulates*
- NFPA 664, *Standard for the Prevention of Fires and Explosions in Wood Processing and Woodworking Facilities*

Most of the NFPA codes pertaining to explosive materials, oxidizers, peroxides, and combustible metals also address deflagration venting. See Section 2 of this chapter for lists of those codes.

Flood, Earthquake, Snow, and Wind Design

Flood, earthquake, snow, and wind design of buildings are all of concern to fire protection engineers because these natural perils can either cause fires or take important fire protection systems out of service. Any of these perils can also slow or prevent response of emergency services to a facility, including firefighting services.

In addition to building codes, many NFPA codes contain design features intended to mitigate the effects of natural perils. Many of the NFPA codes listed in Sections 2 and 4 of this chapter address these perils.

Flooding can cause electrical power loss and disable alarm systems, security systems, process safety controls, and other electrical equipment critical to fire safety. Flooding can also wash away tanks, including tanks holding fire protection water supplies and fuel for engine-driven fire pumps. Therefore, judicious layout and structural design of facilities near flood plains are an important part of facility fire prevention programs.

In addition to the codes listed in Section 2 of this chapter, NFPA codes that address design for protection from flooding include:

- NFPA 1, *Fire Prevention Code*
- NFPA 20, *Standard for the Installation of Stationary Pumps for Fire Protection*
- NFPA 30, *Flammable and Combustible Liquids Code*©
- NFPA 70, *National Electrical Code*©
- NFPA 75, *Standard for the Protection of Electronic Computer/Data Processing Equipment*
- NFPA 99, *Standard for Health Care Facilities*

Many other NFPA codes also contain provisions for protection of buildings, rooms, equipment, and fire protection systems from the effects of flooding.

Proper building structural design for earthquake resistance is essential for achieving earthquake resistance of systems attached to the structure, including piped systems. Earthquakes can break fuel gas piping, which is a very common cause of fires and explosions. The many NFPA codes that cover compressed and liquefied fuel gases have earthquake design provisions.

Earthquakes can also break fire protection piping, taking these systems out of service. If the break cannot be isolated, this reduces the water available for other fire protection systems or for manual firefighting.

Earthquakes can cause many other fire protection-related problems, including breaking flammable liquids storage tanks and tanks holding fire protection water. Damage from earthquakes can also restrict accessibility to the site by emergency responders.

Most codes that cover the installation of piped fire protection systems contain provisions for designing them to withstand earthquakes. NPFA 13, *Standard for the Installation of Sprinkler Systems*, is the most familiar example of a code for piped fire protection systems that contains earthquake design provisions. Other codes containing such provisions include those listed in Section 4 of this chapter.

Wind perils include high winds, monsoons, tornadoes, and hurricanes. Windstorms also bring the potential for water damage, whether it is from wind-driven rain, storm water run-off, or flooding. Wind can down power lines and topple elevated tanks. Fallen trees can damage buildings, process equipment and storage tanks. Like flood and earthquake, windstorms can cause fires and impair fire protection systems.

Building codes contain the structural design features for flood, earthquake and wind, as well as for snow, ice, and other loads. Excessive snow and ice can cause roof collapse, which would break any fire protection piping in the collapsed area.

Most of the building codes currently used in the U.S. already incorporate the provisions of ASCE-7, *Minimum Design Loads for Buildings and Other Structures*. The new editions of the existing building codes and the new national building codes are expected to be adopting ASCE 7-2002.

6:
Fire Protection for New and Existing Buildings

6-1 The Design Process

Buildings are designed to meet many needs. They must have a particular amount of space, accommodate a particular number of occupants, and serve particular functions. They must also have electrical, mechanical, and ventilation systems to support the planned operations. The structural, electrical, and mechanical systems are designed to meet these needs as well as to meet applicable codes. Each discipline is coordinated as necessary to meet the project's goals.

Meeting fire protection goals requires the same type of coordination. Fire protection must be integrated into the design process and coordinated throughout construction. This is true whether planning a new building or making changes to an existing one. As a member of the design team, the fire protection engineer can help identify and resolve fire protection issues and keep them from adversely affecting the project.

Sometimes the fire protection design approach is different for new construction than for existing buildings, but the underlying design principles are the same. Effectively integrating fire protection requires being very familiar with the construction, layout, and occupancy of the building and knowing the functions that the fire protection systems are expected to serve. (See Chapter 2.)

New construction projects are usually more straightforward than projects involving changes to existing buildings. Rather than having to adapt to what is already there, each engineering discipline can start with the designs recognized as being most appropriate.

The optimum design may often be feasible; if not, alternate designs are usually easy to develop.

Making changes to existing buildings can be a design challenge for all the engineering disciplines. Every proposed change can affect all the other disciplines, including fire protection. This is why coordination is very important on projects involving changes to existing buildings.

Sometimes bringing existing buildings up to current prescriptive building codes is not feasible. This is where performance-based fire protection design alternatives can be used to achieve a level of protection equivalent to that specified in the current code. The other engineering disciplines may also use performance-based alternatives to achieve the intent of prescriptive provisions that cannot otherwise be met. In all cases, performance-based alternatives must be coordinated with the other disciplines. And in all cases, these alternatives are best developed by engineers licensed in the appropriate discipline.

Combined architectural/engineering firms are often the prime professionals on very large building projects. Architectural firms are often the prime professionals on moderate sized new construction projects, and engineering consulting firms are often the prime professionals on renovation projects. However, this can vary from project to project. The important thing is for the prime professional to coordinate the designs of all the disciplines, including fire protection, and to help integrate these designs into the project.

6-2 New Construction

While new construction projects may present fewer initial design restrictions, they also present more potential design choices. Making these choices requires research into issues like potential sites, building and process design choices, and other issues. The fire protection engineer would research fire protection water supplies, available alarm services, fire department capabilities, site drainage, fire exposures, special jurisdictional requirements, and other features that may affect fire protection design.

At this point, the fire protection engineer determines whether the usual prescriptive fire protection design best suits the project. This is often the case. The fire protection engineer then implements the prescriptive design while coordinating features of all the relevant codes with the other disciplines.

To implement a performance-based design, the fire protection engineer must know the goals and objectives that have been specified by the owner. All system designs must then be shown to meet these goals and objectives.

The trickiest part of the fire protection design is that it depends on the building construction, layout, and occupancy; and on the mechanical, electrical, and process systems in the building. Therefore, any time any other discipline changes a design, the fire protection has to be checked. This is why fire protection must be coordinated throughout the project.

Coordination is equally important for prescriptive and performance-based designs. In a prescriptive design, a requirement for hose standpipes might apply when the fire area per floor exceeds a particular amount. An initial building design may control these areas by adding fire barriers and doors to corridors on each floor. If a later design revision removes the doors, the hose standpipes would then be required. The fire protection water supply piping to the hose standpipes would need to be sized and placed. This is best done as early as possible in the design.

In a performance-based design, any change to the building can affect whether the fire protection systems meet the performance specifications of the owner. Decreasing the size of an atrium may affect visibility during occupant egress. The smoke control system might have to be redesigned. The amount of combustible materials in a given area can increase when changing construction materials, room layout, occupancy, or storage locations. All such changes would require rechecking the performance of sprinkler systems, if provided, or rechecking the performance of other building systems. Alternately, the configuration of the combustible materials can be changed.

Chapter 4 addresses the coordination issues that accompany performance-based fire protection design. Chapter 5 addresses code coordination issues, which would primarily apply to prescriptive designs. However, such code coordination issues would also apply to performance-based equivalencies; and they might also apply to some performance-based designs.

6-3 Existing Buildings

Even changes seemingly unrelated to fire protection can affect fire protection system design. Moving fire walls can affect the size and value of fire areas, which in turn can affect the required level of protection. Moving unrated walls or partitions can affect local coverage of sprinkler heads or detectors. Changing building layout can affect provisions for means of egress.

Even changes that do not affect the building structure or layout must be considered. Examples include changes in amounts or location of storage, or changes in occupancy. Other examples are changes in interior finishes, process equipment, or building ventilation.

Developing an effective fire protection design requires addressing management of change both during the design process and after completion of the project. Management of change includes many familiar issues, including changes to:

- Building construction;
- Building layout;
- Building systems;
- Occupancy;

- Process design; and
- Storage configuration.

Many projects in existing buildings involve these kinds of changes.

Some building renovation projects could be seen as not involving significant design changes. Such projects can include these types of activities:

- Replacing existing construction materials or building systems with essentially the same materials or systems;
- Using the same as existing type of construction, floor plans, ventilation systems, or protective systems in new areas; or
- Selecting "comparable" or "similar" materials or systems.

These activities might not be seen as changes requiring management. However, managing these activities is important for at least three reasons. First, what a manufacturer deems to be "the same" could have differences of concern in meeting the design goals. Fabrication methods can change, as can specified thicknesses or safety factors. Such things as bolts, filters, gaskets, and other accessories can also change.

Second, even if everything is exactly the same, the question must be asked whether that is appropriate for the new facility. This is a chance to develop new, more relevant facility goals. Third, even if it is appropriate for everything to remain the same, the construction process itself introduces temporary changes that must be addressed. All these factors have the potential of affecting the fire protection design.

The fire protection design process for projects in existing facilities should follow these steps:

- Find and review the existing fire protection design basis.
- If the design basis is not available, analyze existing building construction, building systems, occupancy, and fire protection systems.

- Document the new facility goals.
- Determine any changes required to meet the new goals.
- Analyze the effects of anticipated changes.
- Determine whether prescriptive or performance-based design will best serve the project.
- Consider using combinations of prescriptive and performance-based designs.

Fire protection design can be especially challenging for certain types of renovations to existing buildings. Common examples would be for renovating buildings with restrictive space limitations and for renovating historic buildings while maintaining their character. These types of renovations frequently require performance-based design or performance-based design alternatives to prescriptive code provisions.

NFPA codes relevant to historic buildings include:

- NFPA 909, *Code for the Protection of Cultural Resources*
- NFPA 914, *Code for Fire Protection of Historic Structures*

The following articles from *Fire Protection Engineering* magazine describe how performance-based design or performance-based design alternatives were used in building renovation projects:

Premier Issue (Winter 1999):

> "Fire Protection for the Star Spangled Banner," by Michael J. Rzeznik, P.E. Protecting the fragile fibers in this historic flag required a specially designed combination of fire prevention and fire suppression systems.

Issue Number 2 (Spring 1999):

> "Rehabilitating Existing Buildings," by John M. Watts, Jr., Ph.D. Performance-based design can preserve the character of historic buildings while providing a level of protection comparable to the current code.

Issue Number 5 (Winter 2000):

"Small Atria Smoke Control," by Kurt Ruchala, P.E. A performance-based code equivalency meets the needs of this renovated three-story college dormitory.

Issue Number 8 (Fall 2000):

"Performance-Based Analysis of an Historic Museum," by Andrew Bowman. Renovation of an important historic museum required performance-based analysis of fire safety.

Issue Number 14 (Spring 2002):

"Application of a Systematic Fire Safety Evaluation Procedure in the Protection of Historic Property," by Alexander G. Copping, Ph D. A systematic evaluation procedure is applied to historic churches.

Also see the article "Fire Protection Performance Evaluation for Historic Buildings," by John M. Watts, Jr., *Journal of Fire Protection Engineering*, Vol. 11, No. 4 (2001).

7: Writing Fire Protection Specifications

7-1 Coordinating the Specifications

Project specifications cover all the disciplines on a project. Most specifications have standard divisions for site work and building structural elements, and for mechanical and electrical systems. Other systems, like automated storage and retrieval systems or process control systems, are sometimes given their own divisions for a particular project. Fire protection systems are almost always worked into existing project divisions. Fire protection features covered by project specifications include the following:

- Building construction;
- Building layout;
- Special fire/explosion hazards;
- Detection/alarm systems;
- Fire protection water supplies;
- Sprinkler systems;
- Special extinguishing systems;
- Exposure protection; and
- Fire prevention features.

Other disciplines also cover some of these features, so the project specifications must be coordinated. Chapter 5 lists the fire protection codes, standards, and other references that are relevant to each feature.

Building Construction

This category covers building structural materials and their arrangement, as well as interior finishing materials and their characteristics. It includes the following features:

- Combustibility of structural materials;
- Fire resistance ratings of structural materials;
- Sizes, locations, and ratings of fire areas;
- Protection of openings and penetrations;
- Fireproofing for buildings and other structures;
- Flamespread of interior finishing materials; and
- Amounts and locations of interior finishing materials.

Building Layout

Project specifications should state the underlying assumptions that are incorporated in the building layout and shown on project plans. The building layout must meet the basic requirements for occupant egress and life safety. These assumptions include:

- Building use and occupancy;
- Occupancy classification per applicable codes;
- Hazard classification per applicable codes; and
- Number of occupants to be accommodated.

Special Fire/Explosion Hazards

All buildings have particular fire and/or explosion hazards. Some are common to many buildings, some are characteristic of a particular occupancy, and some are unique to one building. These hazards need to be identified so that appropriate building construction and layout can be planned, and so that appropriate special fire protection systems can be specified. Such hazards include:

- Hazards associated with building utilities;
- Warehousing of ordinary commodities;
- Storage and use of hazardous materials; and
- Hazardous chemical processes.

The hazards associated with building utilities include those associated with fuel firing of heating equipment and with electrically-related hazards. Specifications address these hazards through detailing combustion safeguards and arrangement of electrical systems. The fire loading and chance of extensive downtime presented by grouped cables can also be addressed by specifying where and how they will be run and what protection will be provided for them.

As discussed in Chapter 5, the hazards of materials can include combustibility, flammability, explosivity, or reactivity. They can also include other hazards aggravated by fire, including toxicity, radioactivity, biological activity, and capability for polluting the environment. Hazards of materials are also greatly affected by their configuration. For example, storing products in foam cocoons or pressurizing liquids in aerosol cans increases the hazard. Finally, the hazards of using materials derive from heating, cooling, pressurizing, agitating, or atomizing them.

Detection and Alarm Systems

Detection systems can be associated with fire protection systems, or they can be independent. Both detection and fire protection systems can (and should) be connected to a fire alarm system. Project specifications state what detection and protection will be provided and what parameters the facility's fire alarm system will monitor. System features to be considered include the following:

- Type of automatic detection to be provided in which areas;
- Building fire protection systems to be monitored;
- Monitoring of buildings, equipment, processes, and hazards for off-normal conditions;
- Transmission of alarms to a central alarm receiving station; and
- Evacuation of occupants.

Specifications should include information on any systems that are programmable, including how they are to be programmed and who will have the authority to change the programming. They should also include how the original programming and any changes made to it will be documented.

Another issue that specifications must sometimes address is combined fire alarm and security systems. Today's digital alarm panels can monitor far more conditions than a human operator can assimilate. To assure fast response to fire alarms, the combined panel must be programmed to give fire alarms the highest priority and not allow them to be inactivated or deleted from the display.

Fire Protection Water Supplies

Many features of the fire protection water supply must be specified. The first step is to determine the adequacy and reliability of available fire protection water supplies. Then, the specifications would address how to pipe the supplies to the fire protection systems and whether they need augmenting by pumps, tanks, or alternate supplies. Issues to consider include:

- Location and number of water supplies;
- Location, number, spacing, and type of hydrants;
- Location of sprinkler lead-ins;
- Location of control and sectional valves;
- Location of fire department connections;
- Supplies dedicated to fire protection or for combined use; and
- Need for multiple independent supplies for reliability.

Sprinkler Systems

This is the aspect of fire protection that most architects and engineers are used to specifying. Sprinkler system specifications include:

- Type of pipe and fittings;
- Location, spacing, and type of hangers;
- Location and type of control valves;
- Other valves to be used, such as check valves, alarm check valves, and backflow preventers;
- Type of heads;
- Head spacing;

- System hazard classification;
- System hydraulic design;
- Location of fire department connection; and
- Location of sprinkler riser.

Selecting sprinkler heads is more complex than ever due to the increasing number of varieties now made. This requires knowing the uses and limitations of the heads now available. Specifying the hydraulic design for sprinkler systems is complicated by the fact that these systems might be performance based and do not depend on the prescriptive designs that many engineers have become familiar with. See Chapter 3.

Special Extinguishing Systems

Special extinguishing systems provide additional protection for special fire and explosion hazards when building construction, building layout, and sprinkler system features do not sufficiently mitigate the potential risk. Prime uses of special extinguishing systems are for protecting operations involving flammable or combustible liquids. They have many other uses.

Specifying a special extinguishing system requires determining the following:

- Nature and extent of the hazards at the facility;
- Values at risk from the hazards at the facility;
- Which hazards require special protection;
- Detectors of appropriate responsiveness for the hazard;
- Suitable control system logic for operating the extinguishing system;
- Function of the extinguishing system (fire control, fire extinguishment, exposure protection, inerting, spark suppression, explosion suppression, etc.);
- Extinguishing agents effective for the material that may burn;
- Design basis for the system (rate by volume, rate by area, local application);

- Appropriate hydraulic design for the selected agent and hazard; and
- Suitable storage and delivery system for the agent.

Many types of agents are now used in special extinguishing systems. Fire protection engineers are familiar with how to best apply these agents to particular hazards at a facility.

Exposure Protection

Exposure protection is provided for buildings that may be exposed to fire from outside sources. Sources of exposure fires can include:

- Fire in buildings or structures on the same premises;
- Fire in off-premises buildings or structures;
- Fire or explosion in outside electrical distribution equipment;
- Forest fires or brush fires; and
- Lightning.

Exposure protection can be achieved by providing separation, fireproofing, special equipment design, or water spray systems. Protection from unique perils, like lightning, is achieved with specially designed protective systems.

Fire Prevention Features

Many fire protection features installed in buildings prevent fire from occurring rather than control or extinguish it. Examples of fire prevention features include:

- Combustion safeguards on fuel-fired equipment;
- Heating equipment for hazardous areas;
- Electrical equipment for hazardous areas;
- Diking/drainage for liquids and fire protection water run-off;
- Fail-safe process design;
- Process monitoring and interlocks—pressure, temperature, level, flow, concentration, and other process parameters; and
- Bonding and grounding to control static electricity.

Like fire protection systems, fire prevention features must be specified as part of the building project and coordinated with the other disciplines.

A good example of a fire protection coordination issue that encompasses many disciplines is in sealing fire wall penetrations. The fire protection engineer must confirm this is addressed for power and communications wiring, ductwork, potable water, and process piping, as well as for fire protection piping and wiring. Descriptions of these elements are usually spread through many different sections of the project specifications.

Two primary systems of organizing project specifications are commonly used. Either requires the fire protection specifications to be coordinated. One system uses traditional project specifications and another uses a special section: Division 13 – Special Construction. These next two sections describe these systems.

A new system for organizing project specifications is under development. The last section of this chapter describes the new system.

7-2 Traditional Project Specifications

The most frequently used specifications format is that of the Construction Specifications Institute (CSI). Using a uniform format helps in finding specifications for standard project features on every project. However, this standardization does not usually include all the fire protection aspects of a project.

The Construction Sciences Research Foundation, Inc. (CSRF), a research foundation founded by CSI, developed SPECTEXT® Master Guide Specifications that provide master specifications in CSI format. Many other versions of electronic specifications in CSI format are available. Most Architectural and A/E firms use standard electronic specifications and adapt them to each project.

The CSI format is as follows:

Specifications Division	*Subject*
01	General Requirements
02	Site Construction
03	Concrete
04	Masonry
05	Metals
06	Wood and Plastics
07	Building Protection
08	Doors and Windows
09	Finishes
10	Specialties
11	Equipment
12	Furnishings
13	Special Construction
14	Conveying Systems
15	Mechanical Systems
16	Electrical Systems

The General Requirements division covers site access and safety, construction schedule, and all the project contractual requirements. This division usually includes provisions for substitutions and requirements for acceptance, including tests.

Division 7, Building Protection, is used to specify the building thermal and moisture protection. Division 10, Specialties, includes such items as toilet and bath accessories, louvers, lockers, and folding walls. Division 11, Equipment, includes dock levelers and appliances.

Traditionally, Division 15 contains major sections for potable water and HVAC systems. Sections are also added for other mechanical systems, like water treatment systems, natural gas piping, boilers, chillers, and other mechanical equipment. Sprinkler systems are generally placed in a section in this division, as are inside fire protection standpipes.

To illustrate the coordination required for the fire protection aspects of a project, here is an example of how fire protection specifications may be spread throughout the divisions of project specifications:

Fire Protection Feature	*Applicable Division(s) in Specifications*
Alarm Systems	16
Classified Electrical Equipment	16
Classified Heating Equipment	15
Combustion Safeguards	15
Construction of Buildings	3, 4, 5, 8
Construction of Roofs	7
Detection Systems	16
Elevator Control	14
Emergency Lighting	16
Exit Doors	8
Extinguishers	10
Fire Barrier Penetrations (ducts)	10, 15
Fire Barrier Penetrations (pipes)	15
Fire Barrier Penetrations (wires)	16
Fire Doors	8
Fire Protection Underground	2

Fire Protection Feature	*Applicable Division(s) in Specifications*
Flamespread of Air Handling Filters	15
Flamespread of Insulation	7
Flamespread of Soundproofing	9
Hydrants	2
Interior Finishes	9
Piping, Fittings, Valves, Backflow Preventers, Other Devices	2, 15
Process Safety Control Systems	15, 16, special process division(s)
Smoke Control Systems	15, 16
Special Extinguishing Systems	15
Sprinkler Systems	15
Standpipes	15
Water Supplies (pumps)	2, 3 through 5, 15
Water Supplies (tanks)	2

Two points must be kept in mind concerning the coordination of fire protection features:

1. Fire protection features for particular systems are spread throughout many divisions. For example, the elements of a sprinkler system can be split between divisions 2, 15, and 16.

2. Particular divisions can address several fire protection features. For example, Division 16 can address detectors, alarm systems, smoke control systems, explosion-proof electrical equipment, process safety controls, and protection of wiring penetrations.

7-3 Division 13 – Special Construction

A more recent suggestion made for handling the fire protection systems on a project is to place them in Division 13 – Special Construction. The current CSI draft for this division includes sections as follows:

Division 13 Subsection	*Subject*
13100	Special Facilities and Components (includes kennels, pools, aquariums, and hot tubs)
13200	Storage Tanks (includes elevated, ground level, and underground tanks)
13300	Integrated Construction (includes sound, vibration, and seismic control)
13400	Measurement and Control Instrumentation
13500	Recording Instrumentation
13600	Special Structures (includes air-supported structures, grandstands, greenhouses, observatories, and towers)
13700	Special Purpose Rooms (includes cleanrooms, planetariums, and vaults)
13800	Reserved
13900	Reserved

Some projects have used Section 13800 for fire alarm systems and Section 13900 for sprinkler systems. Section 13200 can be used for tanks storing fire protection water. Process monitoring and interlocks could be placed in Section 13400.

Following the above suggestions does not solve all the coordination issues. For example, specifications for a fire pump would be separated from those for the pump suction tank. Also, many other fire protection-related features would still not be in Division 13, such as fire resistance ratings of walls, fire doors, fire barrier penetrations, interior finishing materials, fire extinguishers, hydrants, smoke control systems, explosion-proof electrical equipment, combustion safeguards on fired equipment, special extinguishing systems, standpipes, and many other features.

On the other hand, if all remaining fire protection and fire prevention features were put in Division 13, the same amount of coordination with the other disciplines would still be required. No one way of handling fire protection specifications eliminates the need for their coordination on a project.

7-4 Expanded Project Specifications

The Construction Specifications Institute is in the process of expanding the MasterFormat™ specifications. New divisions are being created and plans are to move the traditional Division 15 and Division 16 systems to the new divisions.

Following is a summary of the current proposal for expanding specification divisions:

Divisions	*Content*
Divisions 0–2	Procurement, Contracting, General Requirements, and Existing Conditions
Divisions 3–14	Technical Building Construction Divisions, similar to the traditional divisions
Divisions 15–20	Left open for future content
Divisions 21–25	Building Engineering Divisions, including Mechanical, Electrical, Plumbing, and Telecommunications
Divisions 26–30	Left open for future content
Divisions 31–43	Civil Engineering Divisions
Divisions 35–40	Left open for future content
Divisions 41–43	Process Engineering Divisions

As with the other systems of specifications, fire protection features can be spread throughout the divisions. Many items would be included in Divisions 21 through 24. Examples of where various fire protection features could appear are:

Fire Protection Feature	*Applicable Section in Specifications*
Fire Protection Piping/Valves	21300
Fire Pumps	21340
Fire Suppression Systems	21400
Fire Suppression Sprinkler Systems	21410
Fire Extinguishing Systems	21450
Detection and Alarm Systems	21500
Fire Detection and Alarm	21510
Gas Detection and Alarm	21530
Specialty Sensors	21560
Protection Systems	21700
Lightning Protection	21720
Fire and Smoke Protection	21800
Applied Fireproofing	21810
Firestopping	21840
Fire Protection Devices	21900
Fire Extinguishers and Cabinets	21920
Process Piping	22200
Gas Piping	22210
Heat Generation Equipment	23500
Heating Boilers	23510
Fuel-Fired Furnaces	23530
Fuel-Fired Heaters	23540

Fire Protection Feature	*Applicable Section in Specifications*
Facility Electrical Power	24200
Battery Equipment	24240
Lighting	24500
Emergency Lighting	24530

These are just examples of some of the fire protection features that projects could include. There are many others. Divisions 41 through 43 could contain specifications for process safety controls and interlocks that would be important to fire protection.

There is no specific section assigned for smoke control systems. These could be handled under HVAC sections or could be given their own section for a particular project.

The proposed expanded specifications can be viewed on the CSI web site. A place on the site is currently being set up for people to submit comments. The expanded specifications are expected to be made available in late 2003.

References

This chapter lists the references cited in this book, as well as other references relevant to the fire protection aspects of building design. These references include books, handbooks, magazines, journals, building codes, NFPA codes, and UL directories.

The first section lists the related professional organizations with information for contacting them. These organizations published many of the references listed in subsequent sections of this chapter. The subsequent sections list the references alphabetically, by type of reference and by related discipline. The references under related disciplines necessarily overlap, and they may not include every NFPA code that applies to a project for that discipline. Where the codes apply will depend on how the project is coordinated. (See Chapter 7.)

This reference list gives the latest edition for only those references without a specific schedule for updating. Most of the handbooks listed here are updated approximately every five years. Older editions of these handbooks are still useful. Most laboratory listings are updated annually.

Individual NFPA codes are normally updated on three- to five-year cycles, the duration depending on the code. New/updated standards are officially adopted twice a year. Hard copies of the codes are published annually, and electronic codes are published twice a year. Often, building codes refer to a particular edition of an NFPA code. In other cases, the code in effect the year a facility was built may apply. Or, the most recent NFPA code may apply. The NFPA codes listed here are all cited elsewhere in this book. There are many other NFPA codes. Hard copy or electronic sets of all the NFPA codes are available from NFPA.

Related Professional Organizations

Acronym	*Organization and Address*	*Phone/Web Site*
AEI	Architectural Engineering Institute 1801 Alexander Bell Drive Reston, VA 20191-4400	703-295-6360 www.aei.org
AIA	American Institute of Architects 1735 New York Avenue NW Washington, DC 20006	800-242-3837 www.aia.org
AIChE	American Institute of Chemical Engineers 3 Park Avenue New York, New York 10016-5991	212-591-8100 www.aiche.org
ANSI	American National Standards Institute 1819 L Street NW Washington, DC 20036	202-293-8020 www.ansi.org
ASCE	American Society of Civil Engineers 1801 Alexander Bell Drive Reston, VA 20191-4400	703-295-6300 www.asce.org
ASHRAE	American Society of Heating, Refrigerating, and Air-Conditioning Engineers 1791 Tullie Circle NE Atlanta, GA 30329	404-636-8400 www.ashrae.org
ASME	American Society of Mechanical Engineers 1828 L Street NW Washington, DC 20036	202-785-3756 www.asme.org
ASSE	American Society of Safety Engineers 1800 E Oakton Street Des Plaines, IL 60018	847-699-2929 www.asse.org
ASTM	American Society for Testing Materials 100 Barr Harbor Drive PO Box C700 West Conshohocken, PA 19428-2959	610-832-9585 www.astm.org
BOCA	Building Officials and Code Administrators, Inc. 4051 West Flossmoor Road Country Club Hills, IL 60478	708-799-2300 www.bocai.org
CSI	Construction Specifications Institute 99 Canal Center Plaza, Suite 300 Alexandria, VA 22314	800-689-2900 www.csinet.org

Acronym	*Organization and Address*	*Phone/Web Site*
DIERS	Design Institute for Emergency Relief Systems 3 Park Avenue New York, NY 10016-5991	212-591-7353 www.aiche.org/diers
FMRC	FM Research Corporation 1151 Boston-Providence Turnpike PO Box 9102 Norwood, MA 02062	781-762-4300 www.fmglobal.com
ICBO	International Conference of Building Officials 5360 Workman Mill Road Whittier, CA 90601-2298	800-423-6587 www.icbo.org
ICC	International Code Council 5203 Leesburg Pike, Suite 600 Falls Church, VA 22041	702-931-4533 www.intlcode.org
IEEE	Institute of Electrical and Electronics Engineers 1828 L Street NW, Suite 1202 Washington, DC 20036-5104	202-785-0017 www.ieee.org
NCEES	National Council of Examiners for Engineering and Surveying PO Box 1686 Clemson, SC 29633-1686	864-654-6824 www.ncees.org
NFPA	National Fire Protection Association 1 Batterymarch Park Quincy, MA 02269-9101	617-770-3000 www.nfpa.org
NIOSH	National Institute for Occupational Safety and Health 4676 Columbia Parkway Cincinnati, OH 45213	800-356-4674 www.cdc.gov/niosh
NIST	National Institute of Standards and Technology 100 Bureau Drive Gaithersburg, MD 20899	301-975-6478 www.nist.gov
NSPE	National Society of Professional Engineers 1420 King Street Alexandria, VA 22314	703-684-2800 www.nspe.org
OSHA	Occupational Safety & Health Administration 200 Constitution Avenue NW Washington, DC 20210	Six regional offices (see web site) www.osha.gov

Acronym	*Organization and Address*	*Phone/Web Site*
SBCCI	Southern Building Code Congress International 900 Montclair Road Birmingham, AL 35213-1206	205-591-1853 www.sbcci.org
SEI	Structural Engineering Institute 1801 Alexander Bell Drive Reston, VA 20191-4400	703-295-6360 www.seinstitute.org
SFPE	Society of Fire Protection Engineers 7315 Wisconsin Avenue Bethesda, MD 20814	301-718-2910 www.sfpe.org
UL	Underwriters Laboratories 333 Pfingsten Road Northbrook, IL 60062-2096	847-272-8800 www.ul.com

Alphabetical Listing

Approval Guide, published by FM.

Assessing Flame Radiation to External Targets from Pool Fires, published by SFPE.

Automatic Sprinkler Systems Handbook (companion to NFPA 13), published by NFPA.

Building Materials Directory, published by UL.

Chemical Engineers' Handbook, Perry, published by McGraw-Hill, New York.

Combustion and Flame, published by The Combustion Institute, Pittsburgh, PA.

Cracking the Codes: An Architect's Guide to Building Regulations, by Barry D. Yatt, published by John Wiley & Sons, Inc., New York, NY, 1998.

CRC Handbook of Chemistry and Physics, published by CRC Press, Boca Raton, FL.

Dangerous Properties of Industrial Materials, vols. I through III, N. Irving Sax and Richard J. Lewis, Sr., published by Van Nostrand Reinhold, New York, New York.

Design of Smoke Control Systems for Buildings, published by ASHRAE.

Designing Fire Protection for Steel Beams, published by American Iron and Steel Institute, Washington, DC.

Designing Fire Protection for Steel Columns, published by American Iron and Steel Institute, Washington, DC.

Electrical Appliance Utilization Directory, published by UL.

Enclosure Fire Dynamics, by Bjorn Karlsson and James G. Quintiere, published by CRC Press, Boca Raton, FL, 2000.

Eschbach's Handbook of Engineering Fundamentals, published by John Wiley & Sons, Inc., New York, NY.

Fire Alarm Code Handbook (companion to NFPA 72), published by NFPA.
Fire and Materials, published by Wiley Europe Ltd., Chichester, W. Sussex, UK.
Fire Journal, published by NFPA.
Fire Prevention Code Handbook (companion to NFPA 1), published by NFPA.
Fire Protection Engineering, published by SFPE.
Fire Protection Equipment Directory, published by UL.
Fire Protection Guide to Hazardous Materials, published by NFPA.
Fire Protection Handbook, NFPA.
Fire Resistance Directory, volumes I and II, published by UL.
Fire Safety and Loss Prevention, by Kevin Cassidy, published by Butterworth-Heinemann, Woburn, MA, 1992.
Fire Safety Journal, published by Elsevier Science, Woburn, MA.
Fire Safety Science and Engineering, edited by T. Z. Harmathy, published by ASTM Special Technical Publications.
Fire Technology, published jointly by NFPA and Kluwer Academic Publishers.
Flammable and Combustible Liquids Code Handbook (companion to NFPA 30), published by NFPA.
Handbook of Fire Protection Engineering, published by SFPE.
Handbook of Hazardous Chemical Properties, by Nicholas P. Cheremisinoff, published by Butterworth-Heinemann, Woburn, MA.
Hazardous Location Electrical Equipment Directory, published by UL.
Industrial Fire Hazards Handbook, 3d ed., published by NFPA, 1990.
International Building Code, ICC.
Introduction to Fire Dynamics, An, 2d ed., by Dougal Drysdale, published by John Wiley & Sons, Inc., New York, NY, 1998.
Introduction to Performance-Based Fire Safety, by Richard L. P. Custer and Brian J. Meacham, published jointly by SFPE and NFPA, 1997.
Journal of Applied Fire Science, published by Baywood Publishing Company, Amityville, NY.
Journal of Fire Sciences, published by Sage Publications, London, UK.
Journal of Fire Protection Engineering, published by SFPE and Sage Publications, London, UK.
Kirk-Othmer Encyclopedia of Chemical Technology, published by John Wiley & Sons, New York, NY.
Lange's Handbook of Chemistry, published by McGraw-Hill, New York, NY.
Life Safety Code Handbook (companion to NFPA 101), published by NFPA.
Lightning Protection Institute Standard Practice, LPI-175, Lightning Protection Institute, Harvard, IL.
Loss Prevention in the Process Industries, 2d ed., vols. 1 and 2, by Frank P. Lees, published by Butterworth-Heinemann, Woburn, MA, 1996.
LP-Gases Handbook (companion to NFPA 58), published by NFPA.

Marks' Standard Handbook for Mechanical Engineers, published by McGraw-Hill, New York.

Manual of Steel Construction, published by American Institute of Steel Construction, Inc., Chicago, IL.

Minimum Design Loads for Buildings and Other Structures, ASCE-7, published by ASCE.

National Building Code, BOCA.

National Electrical Code Handbook (companion to NFPA 70), published by NFPA.

NFPA Codes:

- 30B, *Aerosol Products, Manufacture and Storage*
- 90A, *Air-Conditioning and Ventilating Systems, Standard for the Installation of*
- 90B, *Air Heating and Air-Conditioning, Standard for the Installation of Warm*
- 61, *Agricultural and Food Products Facilities, Standard for the Prevention of Fires and Dust Explosions in*
- 651, *Aluminum and the Production and Handling of Aluminum Powders, Standard for the Machining and Finishing of*
- 490, *Ammonium Nitrate, Storage of*
- 85, *Boiler and Combustion Systems Hazards Code*
- 251, *Building Construction Materials, Fire Tests of*
- 220, *Building Construction, Types of*
- 255, *Building Materials, Test of Surface Burning Characteristics of*
- 259, *Building Materials, Test Method for Potential Heat of*
- 12, *Carbon Dioxide Extinguishing Systems, Standard on*
- 2001, *Clean Agent Fire Extinguishing Systems, Standard on*
- 318, *Cleanrooms, Protection of*
- 499, *Combustible Dusts and of Hazardous (Classified) Locations for Electrical Installations in Chemical Process Areas, Recommended Practice for the Classification of*
- 654, *Combustible Particulates, Standard for the Prevention of Fire and Dust Explosions from the Manufacturing, Processing, and Handling of*
- 75, *Computer/Data Processing Equipment, Standard for the Protection of Electronic*
- 650, *Conveying Systems for Handling Combustible Particulate Solids, Standard for Pneumatic*
- 96, *Cooking Equipment, Installation of Equipment for the Removal of Smoke and Grease Laden Vapors from Commercial*
- 909, *Cultural Resources, Code for the Protection of*
- 252, *Door Assemblies, Fire Tests of*
- 17, *Dry Chemical Extinguishing Systems, Standard for*
- 70, *Electrical Code,© National*

496, *Electrical Equipment in Hazardous Locations, Purged and Pressurized Enclosures for*
499, *Electrical Installations in Chemical Process Areas, Recommended Practice for the Classification of Combustible Dusts and of Hazardous (Classified) Locations for*
497, *Electrical Installations in Chemical Process Areas, Recommended Practice for the Classification of Flammable Liquids, Gases, or Vapors and of Hazardous (Classified) Locations for*
111, *Emergency and Standby Power, Standard on Stored Electrical Energy*
110, *Emergency and Standby Power Systems, Standard for*
91, *Exhaust Systems for Air Conveying of Vapors, Gases, Mists, and Noncombustible Particulate Solids, Standard for*
69, *Explosion Prevention Systems*
495, *Explosive Materials Code*
80A, *Exposures, Recommended Practice for Protection of Buildings from Exterior Fire*
72, *Fire Alarm Code,© National*
80, *Fire Doors and Fire Windows, Standard for*
10, *Fire Extinguishers, Standard for Portable*
1, *Fire Prevention Code*
24, *Fire Service Mains and Their Appurtenances, Standard for the Installation of Private*
221, *Fire Walls and Fire Barrier Walls*
1124, *Fireworks and Pyrotechnic Articles, Code for the Manufacture, Transportation, and Storage of*
30, *Flammable and Combustible Liquids Code©*
395, *Flammable and Combustible Liquids on Farms and Isolated Construction Projects, Storage of*
33, *Flammable and Combustible Materials, Spray Application of*
497, *Flammable Liquids, Gases, or Vapors and of Hazardous (Classified) Locations for Electrical Installations in Chemical Process Areas, Recommended Practice for the Classification of*
34, *Flammable or Combustible Liquids, Dipping and Coating Processes Using*
253, *Floor Covering Systems Using a Radiant Heat Energy Source, Standard Method of Test for Critical Radiant Flux of*
11, *Foam, Standard for Low Expansion,*
11A, *Foam Systems, Standard for Medium- and High-Expansion*
16, *Foam-Water Sprinkler and Foam-Water Spray Systems, Standard for the Installation of*
86C, *Furnaces Using a Special Processing Atmosphere, Standard for Industrial*

86D, *Furnaces Using Vacuum as an Atmosphere, Standard for Industrial*
54, *Gas Code,© National Fuel*
59A, *Gas, Production, Storage, and Handling of Liquefied Natural*
59, *Gases at Utility Gas Plants, Storage and Handling of Liquefied Petroleum*
55, *Gases in Portable Cylinders, Compressed and Liquefied*
58, *Gases, Storage and Handling of Liquefied Petroleum*
12A, *Halon 1301 Fire Extinguishing Systems, Standard on*
99, *Standard for Health Care Facilities*
271, *Heat and Visible Smoke Release Rates for Materials and Products Using an Oxygen Consumption Calorimeter, Standard Method of Test for*
264, *Heat Release Rates for Materials*
914, *Historic Structures, Code for Fire Protection of*
50A, *Hydrogen Systems at Consumer Sites, Gaseous*
50B, *Hydrogen Systems at Consumer Sites, Liquefied*
99B, *Standard for Hypobaric Facilities*
45, *Laboratories Using Chemicals, Standard on Fire Protection for*
101, *Life Safety Code©*
101A, *Life Safety, Guide on Alternative Approaches to*
780, *Lightning Protection Systems, Standard for the Installation of*
485, *Lithium Metal, Standard for the Storage, Handling, Processing, and Use of*
480, *Magnesium, Storage, Handling, and Processing of*
101B, *Means of Egress for Buildings and Structures*
35, *Organic Coatings, Standard for the Manufacture of*
432, *Organic Peroxide Formulations, Code for the Storage of*
86, *Ovens and Furnaces*
430, *Oxidizers, Storage of Liquid and Solid*
50, *Oxygen Systems at Consumer Sites, Bulk*
53, *Oxygen-Enriched Atmospheres, Fire Hazards in*
51, *Oxygen-Fuel Gas Systems for Welding, Cutting, and Allied Processes, Design and Installation of*
20, *Pumps for Fire Protection, Standard for the Installation of Stationary*
203, *Roof Coverings and Roof Deck Construction, Guide on*
256, *Roof Coverings, Methods of Fire Tests of*
601, *Security Services in Fire Loss Prevention, Standard for*
204, *Smoke and Heat Venting, Guide for*
92B, *Smoke Management Systems in Malls, Atria, and Large Areas, Guide for*
105, *Smoke-Control Door Assemblies, Recommended Practice for the Installation of*

92A, *Smoke-Control Systems, Recommended Practice for*
13, *Sprinkler Systems, Standard for the Installation of*
14, *Standpipe, Private Hydrant, and Hose Systems, Standard for the Installation of*
77, *Static Electricity*
230, *Storage, Standard for the Fire Protection of*
481, *Titanium, Storage, Handling, and Processing of*
68, *Guide for Venting of Deflagrations*
286, *Wall and Ceiling Interior Finish to Room Fire Growth, Standard Methods of Fire Tests for Evaluating Contribution of*
268, *Wall Assemblies Using a Radiant Heat Energy Source, Standard Test Method for Determining Ignitability of Exterior*
265, *Wall Coverings, Evaluating Room Fire Growth Contribution from Textile*
750, *Water Mist Fire Protection Systems, Standard on*
15, *Water Spray Fixed Systems for Fire Protection, Standard for*
22, *Water Tanks for Private Fire Protection, Standard for*
17A, *Wet Chemical Extinguishing Systems, Standard for*
299, *Wildfire, Standard for Protection of Life and Property from*
257, *Window Assemblies, Fire Tests of*
262, *Wires and Cables for Use in Air-Handling Spaces, Standard Method of Test for Flame Travel and Smoke of*
664, *Wood Processing and Woodworking Facilities, Standard for the Prevention of Fires and Explosions in*
482, *Zirconium, Storage, Handling, and Processing of*

NFPA 5000 Building Code, NFPA.
Performance-Based Fire Protection Analysis and Design of Buildings, published by SFPE.
Piloted Ignition of Solid Materials Under Radiant Exposure, SFPE.
Predicting 1st and 2nd Degree Skin Burns, published by SFPE.
Principles of Fire Behavior, by James G. Quintiere, published by Delmar Publishers, Albany, NY, 1998.
Principles of Fire Protection Chemistry and Physics, 3d ed., by Raymond Friedman, published by NFPA, 1998.
Roofing Materials and Systems Directory, published by UL.
Sprinkler Hydraulics, 2d ed., by Harold S. Waas, Jr., published by SFPE, 2001.
Standard Building Code, SBCCI.
Standard for Installation Requirements for Lightning Protection Systems, UL 96A.
Standard for Intrinsically Safe Apparatus and Associated Apparatus for Use in Class I, II, and III, Division 1 Hazardous (Classified) Locations, ANSI/UL 913.
Structural Fire Protection, published by ASCE.
Uniform Building Code, ICBO.

Listing By Type

BOOKS

Cracking the Codes: An Architect's Guide to Building Regulations, by Barry D. Yatt, published by John Wiley & Sons, Inc., New York, NY, 1998.

Design of Smoke Control Systems for Buildings, published by ASHRAE.

Enclosure Fire Dynamics, by Bjorn Karlsson and James G. Quintiere, published by CRC Press, Boca Raton, FL, 2000.

Fire Safety and Loss Prevention, by Kevin Cassidy, published by Butterworth-Heinemann, Woburn, MA, 1992.

Fire Safety Science and Engineering, edited by T. Z. Harmathy, published by ASTM Special Technical Publications.

Introduction to Fire Dynamics, An, 2d ed., by Dougal Drysdale, published by John Wiley & Sons, Inc., New York, NY 1998.

Introduction to Performance-Based Fire Safety, by Richard L. P. Custer and Brian J. Meacham, published jointly by SFPE and NFPA, 1997.

Loss Prevention in the Process Industries, 2d ed., vols. 1 and 2, by Frank P. Lees, published by Butterworth-Heinemann, Woburn, MA, 1996.

Principles of Fire Behavior, by James G. Quintiere, published by Delmar Publishers, Albany, NY, 1998.

Principles of Fire Protection Chemistry and Physics, 3d ed., by Raymond Friedman, published by NFPA, 1998.

Sprinkler Hydraulics, 2d ed., by Harold S. Waas, Jr., published by SFPE, 2001.

Structural Fire Protection, published by ASCE.

BUILDING CODES

International Building Code, ICC.

National Building Code, BOCA.

NFPA 5000 Building Code, NFPA.

Standard Building Code, SBCCI.

Uniform Building Code, ICBO.

HANDBOOKS AND GUIDES

Chemical Engineers' Handbook, Perry, published by McGraw-Hill, New York.

CRC Handbook of Chemistry and Physics, published by CRC Press, Boca Raton, FL.

Dangerous Properties of Industrial Materials, vols. I through III, N. Irving Sax and Richard J. Lewis, Sr., published by Van Nostrand Reinhold, New York, New York.
Designing Fire Protection for Steel Beams, published by American Iron and Steel Institute, Washington, DC.
Designing Fire Protection for Steel Columns, published by American Iron and Steel Institute, Washington, DC.
Eschbach's Handbook of Engineering Fundamentals, published by John Wiley & Sons, Inc., New York, NY.
Handbook of Hazardous Chemical Properties, by Nicholas P. Cheremisinoff, published by Butterworth-Heinemann, Woburn, MA.
Kirk-Othmer Encyclopedia of Chemical Technology, published by John Wiley & Sons, New York, NY.
Lange's Handbook of Chemistry, published by McGraw-Hill, New York.
Lightning Protection Institute Standard Practice, LPI-175, Lightning Protection Institute, Harvard, IL.
Marks' Standard Handbook for Mechanical Engineers, published by McGraw-Hill, New York, NY.
Manual of Steel Construction, published by American Institute of Steel Construction, Inc., Chicago, IL.
Minimum Design Loads for Buildings and Other Structures, ASCE-7, published by ASCE.
NFPA Code Handbooks:
Automatic Sprinkler Systems Handbook (companion to NFPA 13)
Fire Alarm Code Handbook (companion to NFPA 72)
Fire Prevention Code Handbook (companion to NFPA 1)
Flammable and Combustible Liquids Code Handbook (companion to NFPA 30)
Life Safety Code Handbook (companion to NFPA 101)
LP-Gases Handbook (companion to NFPA 58)
National Electrical Code Handbook (companion to NFPA 70)
NFPA Fire Protection Guide to Hazardous Materials.
NFPA Fire Protection Handbook.
NFPA Industrial Fire Hazards Handbook, 3d ed., 1990.
SFPE Engineering Guides:
Assessing Flame Radiation to External Targets from Pool Fires
Performance-Based Fire Protection Analysis and Design of Buildings
Piloted Ignition of Solid Materials Under Radiant Exposure
Predicting 1st and 2nd Degree Skin Burns

SFPE Handbook of Fire Protection Engineering.

LABORATORY LISTINGS AND STANDARDS

ANSI/UL 913, *Standard for Intrinsically Safe Apparatus and Associated Apparatus for Use in Class I, II, and III, Division 1 Hazardous (Classified) Locations*
FM *Approval Guide*
UL 96A, *Standard for Installation Requirements For Lightning Protection Systems*
UL *Building Materials Directory*
UL *Electrical Appliance Utilization Directory*
UL *Fire Protection Equipment Directory*
UL *Fire Resistance Directory*, volumes I and II
UL *Hazardous Location Electrical Equipment Directory*
UL *Roofing Materials and Systems Directory*

MAGAZINES AND JOURNALS

Combustion and Flame, published by The Combustion Institute, Pittsburgh, PA.
Fire and Materials, published by Wiley Europe Ltd., Chichester, W. Sussex, UK.
Fire Journal, published by NFPA.
Fire Protection Engineering, published by SFPE.
Fire Safety Journal, published by Elsevier Science, Woburn, MA.
Fire Technology, published jointly by NFPA and Kluwer Academic Publishers.
Journal of Applied Fire Science, published by Baywood Publishing Company, Amityville, NY.
Journal of Fire Sciences, published by Sage Publications, London, UK.
Journal of Fire Protection Engineering, published by SFPE and Sage Publications, London, UK.

NFPA CODES

1, *Fire Prevention Code*
10, *Standard for Portable Fire Extinguishers*
11, *Standard for Low Expansion Foam*
11A, *Standard for Medium- and High-Expansion Foam Systems*
12, *Standard on Carbon Dioxide Extinguishing Systems*
12A, *Standard on Halon 1301 Fire Extinguishing Systems*
13, *Standard for the Installation of Sprinkler Systems*
14, *Standard for the Installation of Standpipe, Private Hydrant, and Hose Systems*
15, *Standard for Water Spray Fixed Systems for Fire Protection*

16, *Standard for the Installation of Foam-Water Sprinkler and Foam-Water Spray Systems*
17, *Standard for Dry Chemical Extinguishing Systems*
17A, *Standard for Wet Chemical Extinguishing Systems*
20, *Standard for the Installation of Stationary Pumps for Fire Protection*
22, *Standard for Water Tanks for Private Fire Protection*
24, *Standard for the Installation of Private Fire Service Mains and Their Appurtenances*
30, *Flammable and Combustible Liquids Code©*
30B, *Aerosol Products, Manufacture and Storage*
33, *Spray Application of Flammable and Combustible Materials*
34, *Dipping and Coating Processes Using Flammable or Combustible Liquids*
35, *Standard for the Manufacture of Organic Coatings*
45, *Standard on Fire Protection for Laboratories Using Chemicals*
50, *Bulk Oxygen Systems at Consumer Sites*
50A, *Gaseous Hydrogen Systems at Consumer Sites*
50B, *Liquefied Hydrogen Systems at Consumer Sites*
51, *Design and Installation of Oxygen-Fuel Gas Systems for Welding, Cutting, and Allied Processes*
53, *Fire Hazards in Oxygen-Enriched Atmospheres*
54, *National Fuel Gas Code©*
55, *Compressed and Liquefied Gases in Portable Cylinders*
58, *Storage and Handling of Liquefied Petroleum Gases*
59, *Storage and Handling of Liquefied Petroleum Gases at Utility Gas Plants*
59A, *Production, Storage, and Handling of Liquefied Natural Gas*
61, *Standard for the Prevention of Fires and Dust Explosions in Agricultural and Food Products Facilities*
68, *Guide for Venting of Deflagrations*
69, *Explosion Prevention Systems*
70, *National Electrical Code©*
72, *National Fire Alarm Code©*
75, *Standard for the Protection of Electronic Computer/Data Processing Equipment*
77, *Static Electricity*
80, *Standard for Fire Doors and Fire Windows*
80A, *Recommended Practice for Protection of Buildings from Exterior Fire Exposures*
85, *Boiler and Combustion Systems Hazards Code*
86, *Ovens and Furnaces*
86C, *Standard for Industrial Furnaces Using a Special Processing Atmosphere*

86D, *Standard for Industrial Furnaces Using Vacuum as an Atmosphere*
90A, *Standard for the Installation of Air-Conditioning and Ventilating Systems*
90B, *Standard for the Installation of Warm Air Heating and Air-Conditioning*
91, *Standard for Exhaust Systems for Air Conveying of Vapors, Gases, Mists, and Noncombustible Particulate Solids*
92A, *Recommended Practice for Smoke-Control Systems*
92B, *Guide for Smoke Management Systems in Malls, Atria, and Large Areas*
96, *Installation of Equipment for the Removal of Smoke and Grease Laden Vapors from Commercial Cooking Equipment*
99, *Standard for Health Care Facilities*
99B, *Standard for Hypobaric Facilities*
101, *Life Safety Code*©
101A, *Guide on Alternative Approaches to Life Safety*
101B, *Means of Egress for Buildings and Structures*
105, *Recommended Practice for the Installation of Smoke-Control Door Assemblies*
110, *Standard for Emergency and Standby Power Systems*
111, *Standard on Stored Electrical Energy Emergency and Standby Power*
203, *Guide on Roof Coverings and Roof Deck Construction*
204, *Guide for Smoke and Heat Venting*
220, *Types of Building Construction*
221, *Fire Walls and Fire Barrier Walls*
230, *Standard for the Fire Protection of Storage*
251, *Fire Tests of Building Construction Materials*
252, *Fire Tests of Door Assemblies*
253, *Standard Method of Test for Critical Radiant Flux of Floor Covering Systems Using a Radiant Heat Energy Source*
255, *Test of Surface Burning Characteristics of Building Materials*
256, *Methods of Fire Tests of Roof Coverings*
257, *Fire Tests of Window Assemblies*
259, *Test Method for Potential Heat of Building Materials*
262, *Standard Method of Test for Flame Travel and Smoke of Wires and Cables for Use in Air-Handling Spaces*
264, *Heat Release Rates for Materials*
265, *Evaluating Room Fire Growth Contribution from Textile Wall Coverings*
268, *Standard Test Method for Determining Ignitability of Exterior Wall Assemblies Using a Radiant Heat Energy Source*

271, *Standard Method of Test for Heat and Visible Smoke Release Rates for Materials and Products Using an Oxygen Consumption Calorimeter*
286, *Standard Methods of Fire Tests for Evaluating Contribution of Wall and Ceiling Interior Finish to Room Fire Growth*
299, *Standard for Protection of Life and Property from Wildfire*
318, *Protection of Cleanrooms*
395, *Storage of Flammable and Combustible Liquids on Farms and Isolated Construction Projects*
430, *Storage of Liquid and Solid Oxidizers*
432, *Code for the Storage of Organic Peroxide Formulations*
480, *Storage, Handling, and Processing of Magnesium*
481, *Storage, Handling, and Processing of Titanium*
482, *Storage, Handling, and Processing of Zirconium*
485, *Standard for the Storage, Handling, Processing, and Use of Lithium Metal*
490, *Storage of Ammonium Nitrate*
495, *Explosive Materials Code*
496, *Purged and Pressurized Enclosures for Electrical Equipment in Hazardous Locations*
497, *Recommended Practice for the Classification of Flammable Liquids, Gases, or Vapors and of Hazardous (Classified) Locations for Electrical Installations in Chemical Process Areas*
499, *Recommended Practice for the Classification of Combustible Dusts and of Hazardous (Classified) Locations for Electrical Installations in Chemical Process Areas*
601, *Standard for Security Services in Fire Loss Prevention*
650, *Standard for Pneumatic Conveying Systems for Handling Combustible Particulate Solids*
651, *Standard for the Machining and Finishing of Aluminum and the Production and Handling of Aluminum Powders*
654, *Standard for the Prevention of Fire and Dust Explosions from the Manufacturing, Processing, and Handling of Combustible Particulates*
664, *Standard for the Prevention of Fires and Explosions in Wood Processing and Woodworking Facilities*
750, *Standard on Water Mist Fire Protection Systems*
780, *Standard for the Installation of Lightning Protection Systems*
909, *Code for the Protection of Cultural Resources*
914, *Code for Fire Protection of Historic Structures*
1124, *Code for the Manufacture, Transportation, and Storage of Fireworks and Pyrotechnic Articles*
2001, *Standard on Clean Agent Fire Extinguishing Systems*

Listing By Related Discipline

ARCHITECTURAL

Cracking the Codes: An Architect's Guide to Building Regulations, by Barry D. Yatt, published by John Wiley & Sons, Inc., New York, NY, 1998.

Fire Prevention Code Handbook (companion to NFPA 1), NFPA.

International Building Code, ICC.

Life Safety Code Handbook (companion to NFPA 101), NFPA.

National Building Code, BOCA.

NFPA Codes:

- 1, *Fire Prevention Code*
- 10, *Standard for Portable Fire Extinguishers*
- 14, *Standard for the Installation of Standpipe, Private Hydrant, and Hose Systems*
- 80A, *Recommended Practice for Protection of Buildings from Exterior Fire Exposures*
- 101, *Life Safety Code*©
- 101A, *Guide on Alternative Approaches to Life Safety*
- 101B, *Means of Egress for Buildings and Structures*
- 220, *Types of Building Construction*
- 253, *Standard Method of Test for Critical Radiant Flux of Floor Covering Systems Using a Radiant Heat Energy Source*
- 255, *Test of Surface Burning Characteristics of Building Materials*
- 264, *Heat Release Rates for Materials*
- 265, *Evaluating Room Fire Growth Contribution from Textile Wall Coverings*
- 286, *Standard Methods of Fire Tests for Evaluating Contribution of Wall and Ceiling Interior Finish to Room Fire Growth*
- 299, *Standard for Protection of Life and Property from Wildfire*
- 601, *Standard for Security Services in Fire Loss Prevention*
- 780, *Standard for the Installation of Lightning Protection Systems*

NFPA 5000 Building Code, NFPA.

Standard Building Code, SBCCI.

Uniform Building Code, ICBO.

CHEMICAL

Chemical Engineers' Handbook, Perry, published by McGraw-Hill, New York.

CRC Handbook of Chemistry and Physics, published by CRC Press, Boca Raton, FL.

Dangerous Properties of Industrial Materials, vols. I through III, N. Irving Sax and Richard J. Lewis, Sr., published by Van Nostrand Reinhold, New York, New York.

Flammable and Combustible Liquids Code Handbook (companion to NFPA 30), NFPA.

Fire Protection Guide to Hazardous Materials, NFPA.

Handbook of Hazardous Chemical Properties, by Nicholas P. Cheremisinoff, published by Butterworth-Heinemann, Woburn, MA.

Kirk-Othmer Encyclopedia of Chemical Technology, published by John Wiley & Sons, New York, NY.

Lange's Handbook of Chemistry, published by McGraw-Hill, New York.

Loss Prevention in the Process Industries, 2d ed., vols. 1 and 2, by Frank P. Lees, published by Butterworth-Heinemann, Woburn, MA, 1996.

LP-Gases Handbook (companion to NFPA 58), NFPA.

NFPA Codes:

- 30, *Flammable and Combustible Liquids Code*©
- 30B, *Aerosol Products, Manufacture and Storage*
- 33, *Spray Application of Flammable and Combustible Materials*
- 34, *Dipping and Coating Processes Using Flammable or Combustible Liquids*
- 35, *Standard for the Manufacture of Organic Coatings*
- 45, *Standard on Fire Protection for Laboratories Using Chemicals*
- 50, *Bulk Oxygen Systems at Consumer Sites*
- 50A, *Gaseous Hydrogen Systems at Consumer Sites*
- 50B, *Liquefied Hydrogen Systems at Consumer Sites*
- 51, *Design and Installation of Oxygen-Fuel Gas Systems for Welding, Cutting, and Allied Processes*
- 53, *Fire Hazards in Oxygen-Enriched Atmospheres*
- 54, *National Fuel Gas Code*©
- 55, *Compressed and Liquefied Gases in Portable Cylinders*
- 58, *Storage and Handling of Liquefied Petroleum Gases*
- 59, *Storage and Handling of Liquefied Petroleum Gases at Utility Gas Plants*
- 59A, *Production, Storage, and Handling of Liquefied Natural Gas*
- 61, *Standard for the Prevention of Fires and Dust Explosions in Agricultural and Food Products Facilities*

68, *Guide for Venting of Deflagrations*
69, *Explosion Prevention Systems*
77, *Static Electricity*
86, *Ovens and Furnaces*
86C, *Standard for Industrial Furnaces Using a Special Processing Atmosphere*
86D, *Standard for Industrial Furnaces Using Vacuum as an Atmosphere*
96, *Installation of Equipment for the Removal of Smoke and Grease Laden Vapors from Commercial Cooking Equipment*
99B, *Standard for Hypobaric Facilities*
318, *Protection of Cleanrooms*
395, *Storage of Flammable and Combustible Liquids on Farms and Isolated Construction Projects*
430, *Storage of Liquid and Solid Oxidizers*
432, *Code for the Storage of Organic Peroxide Formulations*
480, *Storage, Handling, and Processing of Magnesium*
481, *Storage, Handling, and Processing of Titanium*
482, *Storage, Handling, and Processing of Zirconium*
485, *Standard for the Storage, Handling, Processing, and Use of Lithium Metal*
490, *Storage of Ammonium Nitrate*
495, *Explosive Materials Code*
496, *Purged and Pressurized Enclosures for Electrical Equipment in Hazardous Locations*
497, *Recommended Practice for the Classification of Flammable Liquids, Gases, or Vapors and of Hazardous (Classified) Locations for Electrical Installations in Chemical Process Areas*
499, *Recommended Practice for the Classification of Combustible Dusts and of Hazardous (Classified) Locations for Electrical Installations in Chemical Process Areas*
651, *Standard for the Machining and Finishing of Aluminum and the Production and Handling of Aluminum Powders*
664, *Standard for the Prevention of Fires and Explosions in Wood Processing and Woodworking Facilities*
1124, *Code for the Manufacture, Transportation, and Storage of Fireworks and Pyrotechnic Articles*

Principles of Fire Protection Chemistry and Physics, 3d ed., by Raymond Friedman, published by NFPA, 1998.

ELECTRICAL

Electrical Appliance Utilization Directory, UL.
Fire Alarm Code Handbook (companion to NFPA 72), NFPA.
Hazardous Location Electrical Equipment Directory, UL.
Lightning Protection Institute Standard Practice, LPI-175, Lightning Protection Institute, Harvard, IL.
National Electrical Code Handbook (companion to NFPA 70), NFPA.
NFPA Codes:
11, *Standard for Low Expansion Foam*
11A, *Standard for Medium- and High-Expansion Foam Systems*
12, *Standard on Carbon Dioxide Extinguishing Systems*
12A, *Standard on Halon 1301 Fire Extinguishing Systems*
15, *Standard for Water Spray Fixed Systems for Fire Protection*
16, *Standard for the Installation of Foam-Water Sprinkler and Foam-Water Spray Systems*
17, *Standard for Dry Chemical Extinguishing Systems*
17A, *Standard for Wet Chemical Extinguishing Systems*
20, *Standard for the Installation of Stationary Pumps for Fire Protection*
70, *National Electrical Code©*
72, *National Fire Alarm Code©*
77, *Static Electricity*
101, *Life Safety Code©*
101A, *Guide on Alternative Approaches to Life Safety*
101B, *Means of Egress for Buildings and Structures*
110, *Standard for Emergency and Standby Power Systems*
111, *Standard on Stored Electrical Energy Emergency and Standby Power*
262, *Standard Method of Test for Flame Travel and Smoke of Wires and Cables for Use in Air-Handling Spaces*
497, *Recommended Practice for the Classification of Flammable Liquids, Gases, or Vapors and of Hazardous (Classified) Locations for Electrical Installations in Chemical Process Areas*
499, *Recommended Practice for the Classification of Combustible Dusts and of Hazardous (Classified) Locations for Electrical Installations in Chemical Process Areas*
750, *Standard on Water Mist Fire Protection Systems*
2001, *Standard on Clean Agent Fire Extinguishing Systems*

Standard for Installation Requirements for Lightning Protection Systems, UL 96A.
Standard for Intrinsically Safe Apparatus and Associated Apparatus for Use in Class I, II, and III, Division 1 Hazardous (Classified) Locations, ANSI/UL 913.

FIRE SCIENCE

Assessing Flame Radiation to External Targets from Pool Fires, SFPE.
Combustion and Flame, published by The Combustion Institute, Pittsburgh, PA.
Dangerous Properties of Industrial Materials, vols. I through III, N. Irving Sax and Richard J. Lewis, Sr., published by Van Nostrand Reinhold, New York, New York.
Enclosure Fire Dynamics, by Bjorn Karlsson and James G. Quintiere, published by CRC Press, Boca Raton, FL, 2000.
Fire and Materials, published by Wiley Europe Ltd., Chichester, W. Sussex, UK.
Fire Journal, published by NFPA.
Fire Protection Engineering, published by SFPE.
Fire Protection Guide to Hazardous Materials, NFPA.
Fire Protection Handbook, NFPA.
Fire Safety and Loss Prevention, by Kevin Cassidy, published by Butterworth-Heinemann, Woburn, MA, 1992.
Fire Safety Journal, published by Elsevier Science, Woburn, MA.
Fire Safety Science and Engineering, edited by T. Z. Harmathy, published by ASTM Special Technical Publications.
Fire Technology, published jointly by NFPA and Kluwer Academic Publishers.
Handbook of Fire Protection Engineering, SFPE.
Handbook of Hazardous Chemical Properties, by Nicholas P. Cheremisinoff, published by Butterworth-Heinemann, Woburn, MA.
Industrial Fire Hazards Handbook, 3d ed., 1990, NFPA.
Introduction to Fire Dynamics, An, 2d ed., by Dougal Drysdale, published by John Wiley & Sons, Inc., New York, NY 1998.
Introduction to Performance-Based Fire Safety, by Richard L. P. Custer and Brian J. Meacham, published jointly by SFPE and NFPA, 1997.
Journal of Applied Fire Science, published by Baywood Publishing Company, Amityville, NY.
Journal of Fire Sciences, published by Sage Publications, London, UK.

Journal of Fire Protection Engineering, published by SFPE and Sage Publications, London, UK.
Loss Prevention in the Process Industries, 2d ed., vols. 1 and 2, by Frank P. Lees, published by Butterworth-Heinemann, Woburn, MA, 1996.
Performance-Based Fire Protection Analysis and Design of Buildings, SFPE.
Piloted Ignition of Solid Materials Under Radiant Exposure, SFPE.
Predicting 1st and 2nd Degree Skin Burns, SFPE.
Principles of Fire Behavior, by James G. Quintiere, published by Delmar Publishers, Albany, NY, 1998.
Principles of Fire Protection Chemistry and Physics, 3d ed., by Raymond Friedman, published by NFPA, 1998.

MECHANICAL

Automatic Sprinkler Systems Handbook (companion to NFPA 13), NFPA.
Design of Smoke Control Systems for Buildings, published by ASHRAE.
Fire Protection Equipment Directory, UL.
NFPA Codes:
11, *Standard for Low Expansion Foam*
11A, *Standard for Medium- and High-Expansion Foam Systems*
12, *Standard on Carbon Dioxide Extinguishing Systems*
12A, *Standard on Halon 1301 Fire Extinguishing Systems*
13, *Standard for the Installation of Sprinkler Systems*
14, *Standard for the Installation of Standpipe, Private Hydrant, and Hose Systems*
15, *Standard for Water Spray Fixed Systems for Fire Protection*
16, *Standard for the Installation of Foam-Water Sprinkler and Foam-Water Spray Systems*
17, *Standard for Dry Chemical Extinguishing Systems*
17A, *Standard for Wet Chemical Extinguishing Systems*
20, *Standard for the Installation of Stationary Pumps for Fire Protection*
22, *Standard for Water Tanks for Private Fire Protection*
24, *Standard for the Installation of Private Fire Service Mains and Their Appurtenances*
30, *Flammable and Combustible Liquids Code*©
45, *Standard on Fire Protection for Laboratories Using Chemicals*
68, *Guide for Venting of Deflagrations*

69, *Explosion Prevention Systems*
85, *Boiler and Combustion Systems Hazards Code*
86, *Ovens and Furnaces*
86C, *Standard for Industrial Furnaces Using a Special Processing Atmosphere*
86D, *Standard for Industrial Furnaces Using Vacuum as an Atmosphere*
90A, *Standard for the Installation of Air-Conditioning and Ventilating Systems*
90B, *Standard for the Installation of Warm Air Heating and Air-Conditioning*
91, *Standard for Exhaust Systems for Air Conveying of Vapors, Gases, Mists, and Noncombustible Particulate Solids*
92A, *Recommended Practice for Smoke-Control Systems*
92B, *Guide for Smoke Management Systems in Malls, Atria, and Large Areas*
96, *Installation of Equipment for the Removal of Smoke and Grease Laden Vapors from Commercial Cooking Equipment*
204, *Guide for Smoke and Heat Venting*
262, *Standard Method of Test for Flame Travel and Smoke of Wires and Cables for Use in Air-Handling Spaces*
271, *Standard Method of Test for Heat and Visible Smoke Release Rates for Materials and Products Using an Oxygen Consumption Calorimeter*
318, *Protection of Cleanrooms*
650, *Standard for Pneumatic Conveying Systems for Handling Combustible Particulate Solids*
750, *Standard on Water Mist Fire Protection Systems*
2001, *Standard on Clean Agent Fire Extinguishing Systems*

Marks' Standard Handbook for Mechanical Engineers, published by McGraw-Hill, New York.

Sprinkler Hydraulics, 2d ed., by Harold S. Waas, Jr., published by SFPE, 2001.

STRUCTURAL

Building Materials Directory, UL.

Designing Fire Protection for Steel Beams, published by American Iron and Steel Institute, Washington, DC.

Designing Fire Protection for Steel Columns, published by American Iron and Steel Institute, Washington, DC.

Fire Prevention Code Handbook (companion to NFPA 1), NFPA.

Fire Resistance Directory, UL.

International Building Code, ICC.
Life Safety Code Handbook (companion to NFPA 101), NFPA.
Manual of Steel Construction, published by American Institute of Steel Construction, Inc., Chicago, IL.
Minimum Design Loads for Buildings and Other Structures, ASCE-7, published by ASCE.
National Building Code, BOCA.
NFPA Codes:
1, *Fire Prevention Code*
15, *Standard for Water Spray Fixed Systems for Fire Protection*
30, *Flammable and Combustible Liquids Code*©
30B, *Aerosol Products, Manufacture and Storage*
33, *Spray Application of Flammable and Combustible Materials*
35, *Standard for the Manufacture of Organic Coatings*
45, *Standard on Fire Protection for Laboratories Using Chemicals*
61, *Standard for the Prevention of Fires and Dust Explosions in Agricultural and Food Products Facilities*
68, *Guide for Venting of Deflagrations*
80, *Standard for Fire Doors and Fire Windows*
86, *Ovens and Furnaces*
91, *Standard for Exhaust Systems for Air Conveying of Vapors, Gases, Mists, and Noncombustible Particulate Solids*
101, *Life Safety Code*©
101A, *Guide on Alternative Approaches to Life Safety*
101B, *Means of Egress for Buildings and Structures*
203, *Guide on Roof Coverings and Roof Deck Construction*
220, *Types of Building Construction*
221, *Fire Walls and Fire Barrier Walls*
251, *Fire Tests of Building Construction Materials*
252, *Fire Tests of Door Assemblies*
256, *Methods of Fire Tests of Roof Coverings*
257, *Fire Tests of Window Assemblies*
259, *Test Method for Potential Heat of Building Materials*
268, *Standard Test Method for Determining Ignitability of Exterior Wall Assemblies Using a Radiant Heat Energy Source*
318, *Protection of Cleanrooms*
654, *Standard for the Prevention of Fire and Dust Explosions from the Manufacturing, Processing, and Handling of Combustible Particulates*

664, *Standard for the Prevention of Fires and Explosions in Wood Processing and Woodworking Facilities*
NFPA 5000 Building Code, NFPA.
Roofing Materials and Systems Directory, UL.
Standard Building Code, SBCCI.
Structural Fire Protection, published by ASCE.
Uniform Building Code, ICBO.

Index

About the Author

Jane Lataille has worked in the field of fire protection engineering since 1976. Twenty-four years of that work was with Industrial Risk Insurers (now GE Global Asset Protection Services), and two years involved preparing fire protection plans and specifications for a multidiscipline engineering consulting firm. She is now a fire protection engineer with the Los Alamos National Laboratory.

Jane is very active with the Society of Fire Protection Engineers (SFPE), having served on the Board of Directors and as Engineering Licensing Committee Chair, and staying active on the Publications and Engineering Licensing Committees. She helped start SFPE's newest magazine, *Fire Protection Engineering*, and serves on its Editorial Advisory Board. Throughout the years, SFPE has acknowledged Jane's support of the fire protection engineering discipline with the Hat's Off Award, the President's Award, and most recently, the D. Peter Lund Award in May 2002. SFPE elected Jane to the grade of Fellow in 1997.

She has also been active with the National Fire Protection Association (NFPA), serving on 12 committees responsible for preparing over 20 codes and standards. She has published several chapters in the last four editions of the *NFPA Fire Protection Handbook* and one chapter in the new NFPA handbook *Understanding Flammable Liquids*. She is currently managing NFPA's new handbook on *Fire Protection of Storage Facilities*.

Jane received a BS in Physics from Worcester Polytechnic Institute and an MS in Electrical Engineering from Rensselaer Polytechnic Institute. She also earned the Associate in Risk Management designation from the Insurance Institute of America. She is a licensed Professional Engineer in five states.